T0122772

The Rise of Horses

The Rise of Horses

55 MILLION YEARS OF EVOLUTION

Jens Lorenz Franzen

Translated by Kirsten M. Brown

THE JOHNS HOPKINS UNIVERSITY PRESS BALTIMORE

Originally published as *Die Urpferde der Morgenröte: Ursprung und Evolution der Pferde.* © 2006 Elsevier GmbH, Spektrum Akademischer Verlag, Heidelberg. ISBN 978-3-8274-1680-3.

Jens Lorenz Franzen is the author and retains the moral right to be named as author in all publications of this volume. The English translation was undertaken by the Johns Hopkins University Press.

Printed in China on acid-free paper
9 8 7 6 5 4 3 2 1

The Johns Hopkins University Press
2715 North Charles Street
Baltimore, Maryland 21218-4363
www.press.jhu.edu

Library of Congress Cataloging-in-Publication Data

Franzen, Jens Lorenz.
[Urpferde der Morgenröte. English]
The rise of horses : 55 million years of evolution / Jens Lorenz Franzen ; translated by Kirsten M. Brown.
p. cm.
Includes bibliographical references and index.
ISBN-13: 978-0-8018-9373-5 (hardcover : alk. paper)
ISBN-10: 0-8018-9373-9 (hardcover : alk. paper)
1. Horses—Evolution. 2. Horses, Fossil. 3. Horses—History. I. Title.
QL737.U62F73 2009
599.665'138—dc22 2008056006

A catalog record for this book is available from the British Library.

Illustration credits may be found at the end of this book.

Special discounts are available for bulk purchases of this book. For more information, please contact Special Sales at 410-516-6936 or specialsales@press.jhu.edu.

The Johns Hopkins University Press uses environmentally friendly book materials, including recycled text paper that is composed of at least 30 percent post-consumer waste, whenever possible. All of our book papers are acid-free, and our jackets and covers are printed on paper with recycled content.

CONTENTS

PREFACE TO THE ENGLISH EDITION

It is often the case that books are translated from English into German; less frequently are German books translated into English. This book is one of those German to English cases and there are two main reasons, I think, why my book has found its way into the hands of an American publisher.

The first reason is that there are four famous sites with fossil horses in Germany. One is Grube Messel, which is designated as a World Natural Heritage Site and contains fossils from the Middle Eocene, some 47 million years ago. No fewer than 60 skeletons of early (dawn) horses have been excavated from this former oil shale mine, which is situated not far from popular Frankfurt (am Main) Airport. The fossils at Messel are remarkable, with even the outlines of the soft body visible in several fossils—including the tips of the hairs! From these finds we know, for example, what the outer ears of Eocene horses looked like. Messel is also the first fossil site in the world where gut contents of dawn horses were found—in most cases, leaf fragments but sometimes also seeds. These Messel fossils suggest that William Diller Matthew was correct when, in 1926, he postulated that horses were originally browsers.

The Eckfeld site in the Eifel Mountains, west of Bonn, is somewhat younger than Messel. This site, in the sediments of a former crater lake (similar to Messel), dates to about 44 million years ago. Here too skeletons of dawn horses were found, including one of a pregnant mare with her fetus still wrapped in the remains of the placenta. In eastern Germany is the third site, Geiseltal near Halle (Saale), not far south of Berlin. Here a former lignite mine produced not only coal but also thousands of fossil vertebrates including one complete articulated skeleton of a dawn horse, as well as several fragmentary skeletons, numerous isolated crania, mandibles, teeth, and bones. Fossils from these sites range from about 43 to 47 million years ago. Thus it is possible to follow the course of evolution over a geologically significant time interval, a time when the world experienced the rise of horses.

Quite different in age is a fourth location with fossil horses of unique quality. This is the Höwenegg near Lake Constance in southwest Germany, dating to the Late Miocene. This site is in a former lake at the foot of a volcano that was active about 10 million years ago. The Höwenegg is the only fossil site in the world where complete skeletons of hipparions, in this case *Hippotherium primigenium*, have been excavated, including a pregnant mare with her fetus. The four sites, taken together, provide some of the best fossil evidence concerning the evolution of horses. I hasten to add that I do not ignore the rest of the world. The many important discoveries of early horses from North America are discussed at some length because North America is in fact where most of the evolution of horses took place.

A second reason the book is worthy of translation into English is that it provides an explanation of the so-called Frankfurt theory of evolution (FTE) and its consequences for interpreting the fossil record. Like all organisms, horses are considered to be energy-transforming constructions. Over evolutionary time their energy balance (input versus output) is improved. It is not so much the environment but rather the body construction that restricts and directs evolutionary development. The reader will be fascinated to see why the lateral toes of horses became reduced during millions of years of evolution and how the horses arrived, finally, at a manner of walking, trotting, and even galloping on the tips of the central toes, which are their only remaining toes.

Horses are often used as the classic example of evolution, and I trust that my book reveals why this is so. I am happy to have the opportunity to introduce these amazing animals to a new readership.

The Rise of Horses

CHAPTER ONE

Prologue

THE FIRST HORSE THAT I KNEW PERSONALLY, so to speak, was called Fanny. She was a mare that worked on the farm of a brewery in Bavaria. I got to know Fanny in 1943 when my parents and I left our home in Bremen because of the increasing air raids of World War II. We traveled south, where I met Fanny and Pierre, a French POW. The three of us began working in the fields. Sadly, this first love would have a tragic end. One winter day, Fanny slid on a snow-covered street, broke her leg, and had to be put down. For a long time there were large red stains in the white snow by the side of the road, and for a long time I steered clear of that area.

My next encounter with horses occurred a year later. The war had shifted us to a farm in Lower Saxony, south of Bremen. Working in the field and in the yard, I realized that, like us, horses have distinct personalities and characters. Susi was very headstrong, even dangerous for us children. We knew when to stay out of her way. Lotti, on the other hand, was decidedly gentle. When the horses came home in the evenings tired from fieldwork, we youngsters were allowed to ride them to the meadows for grazing. Naturally, the horses knew the way quite well. They did not need any guidance, but rather accepted their young riders out of sheer good-naturedness. We were buddies. We rode without saddles and without bridles, holding tight to the horse's mane. One night we decided to urge the horses to a faster pace. They took the call seriously. We exploded at full gallop to our school, to the amazement of the villagers, then bolted from there on the fastest path back to the stables. This performance was far more than we had bargained for. Clinging precariously to the horses' manes, we were happy when the adventure ended without injury in front of the homely stables. To our relief, no one on the farm had noticed our escapade. The horses, however, had showed us a lesson.

When the war was over, I returned with my family to Bremen. Long years followed without any horses. The postwar years were anything but easy, and my family could not afford riding lessons. After high school, I decided to study geology and paleontology. It was my doctoral thesis that brought me back to horses, although in the form of their extinct relatives, the palaeotheres.

After obaining my Ph.D. and finishing a short assistantship at the Institute of Geology and Paleontology at Freiberg University, I moved to the famous Forschungsinstitut und Naturmuseum Senckenberg in Frankfurt. I was already at the Naturkundemuseum in Basel, working on my dissertation on palaeotheres, when I became interested in the dawn horses from the Messel Pit. Immediately after my arrival in Frankfurt, I asked what was going on at Messel, in order to explore the possibilities of doing fieldwork there. This was impossible at the time because of a 1912 agreement between the mine operator and the Hessisches Landesmuseum in Darmstadt. The latter owned the sole rights to all fossil findings from the location. However when the oil-shale mining ceased in the early 1970s, plans were made to turn the mine into a giant trash dump for the entire Rhine-Main region. This afforded the opportunity to undertake excavations in the Messel Pit.

During the mining years, only a few more or less black and unspectacular fragments of fossil vertebrates were known from the Messel Pit (Figure 1.1). It was not even known from which corner of the pit they came. In a way, the findings were the by-products of the mining. They were preserved in tubs until a scientist from the Hessisches Landesmuseum in Darmstadt came along. This situation changed only with the end of mining activities. Although the search for fossils in the abandoned surface mines was still not allowed, it was not expressly forbidden. So little by little, small groups of private collectors crossed over the symbolic fences. They went to work with pickaxes and shovels in the dark oil slate.

The activities of the private collectors showed that, contrary to all assumptions, the excavations were worth the effort. In addition, the amateurs applied and further developed a preparation method in which the original fossils are transferred to synthetic resin layers. Slowly the real quality and significance of the Messel fossils became

Figure 1.1. Dawn horse of the species *Eurohippus messelensis* discovered when the mining was still carried out at the Messel Pit near Darmstadt. At that time, findings were carefully dried and soaked with paraffin wax; sometimes they were also embedded on a slab of a wax-chalk mixture.

evident. There were complete skeletons of birds and bats, and the birds' plumage was frequently preserved. There were also several incredibly well-preserved feathers (Figure 1.2), while many bat fossils included skin and hair (Figure 1.3). Since the first discovery of crocodile osteoderms in 1875, the Messel Pit had kept valuable secrets for 100 years. Now a treasure waited to be uncovered.

A find described by the media outlets as the "Messel primitive horse" stirred up great interest. The skeleton had been offered to several large museums around the world for varying sums. Back in Senckenberg we received an anonymous phone call in March 1974. In case we were interested, we could inspect the world-famous find in a bank in southern Frankfurt. I was chosen from our institute to keep this appointment. Armed with two cameras, I hit the road. In the specified bank, no one knew anything. I drove to a branch nearby. However, nothing was known there either. I came back to the original appointed location. Just as I arrived at the bank, a door opened and four men dragged a very heavy object, with blankets cloaking it, into the room. The bank clerks retreated while I went to meet the people,

Figure 1.2. Fossilized bird feathers (about 5 cm long) from the 1975 Senckenberg excavation at the Messel Pit. The feather already shows the typical modern structure: from shaft (rachis), branches (ramii) go off on both sides, bearing small rays (radii) with tiny hooks (radioli) that are joining with each other to form the so-called flag of the feather.

because I suspected that they were carrying what I had come to see. Gruffly I was rebuffed by one of the men: "No cameras, no photos!"

Bit by bit the situation resolved itself. The four men talked with the banker, and about half an hour later they signaled to me that I could now survey the object. In a back room, the mysterious thing was standing on an easel. The blanket was pulled back. I was speechless. Before me was one of the most wonderful fossils I had ever seen (Figure 1.4). Bronze colors glittered against a dark gray background, revealing a complete, fantastically preserved skeleton of a dawn horse. Not until years later did I succeed in getting a cast that showed the discovery was, in reality, a *Hyrachyus,* an early relative of rhinos and tapirs. The sum these people were asking for the specimen was 800,000 DM, minus a 50,000 DM discount for the Senckenberg Museum. I gulped. The discovery was stunning, but so was the price. When I asked once again to take a picture, I was emphatically told that I could neither take a photo nor acquire a cast. My indignation and disappointment grew. Here I stood in front of one of the finest fossils in the entire world, and contrary to what I was accustomed to

Figure 1.3. Fossil bat (*Palaeochiropteryx tupaiodon*) from the Messel Pit, displaying the wing (patagium), the soft body contour, and even an outer ear. In the central body area, the bones have completely dissolved. There is not yet a convincing explanation for this. Acids, which originate with the microbial decomposition of the body, presumably play a role. This object has not yet been transferred into a synthetic resin. The split of the oil shale produced two mirror images that can be prepared separately.

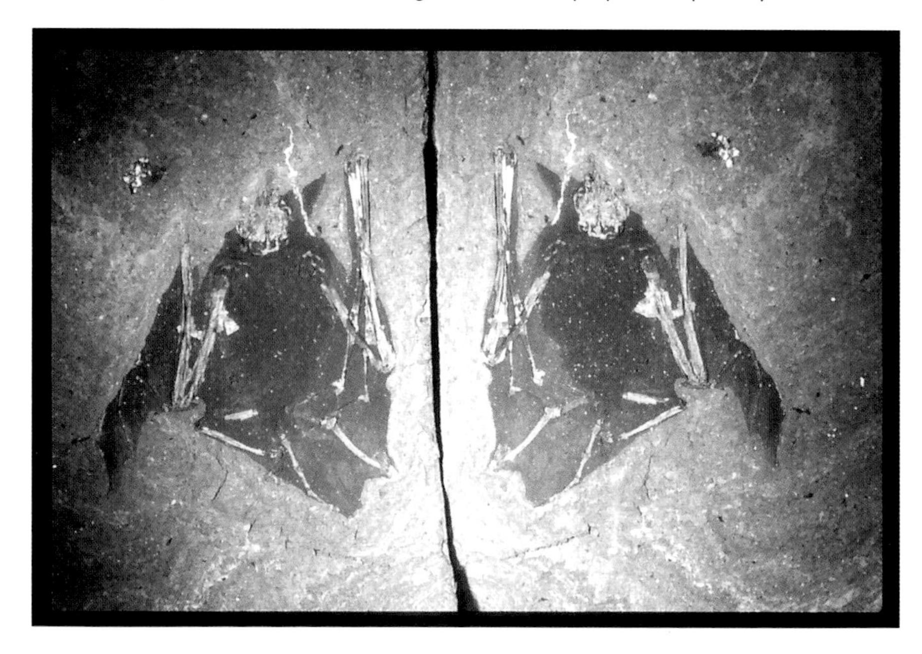

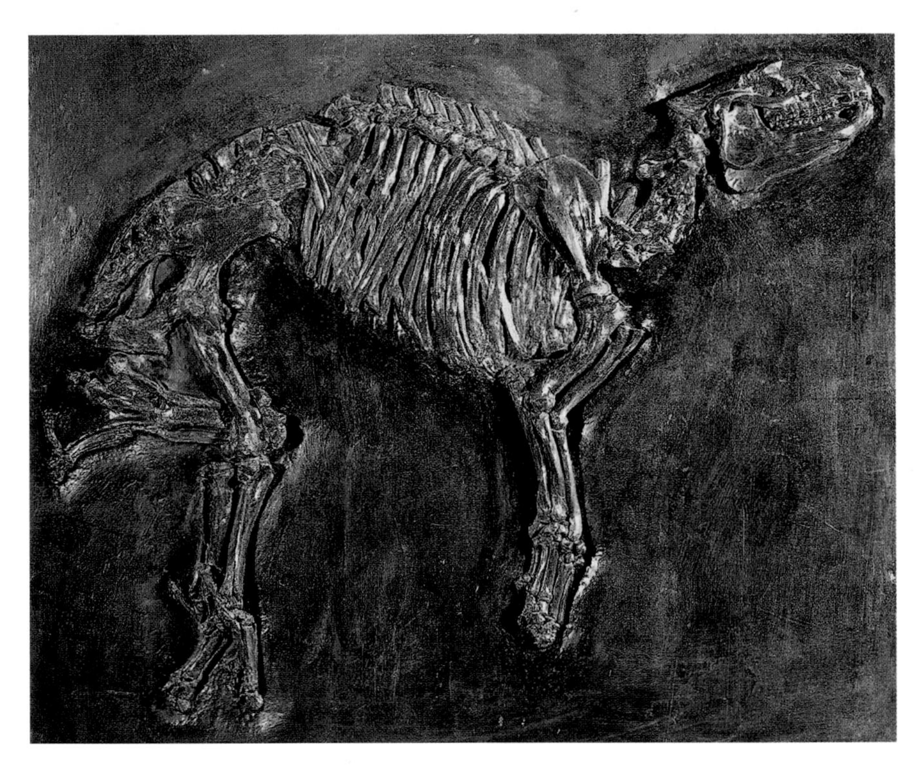

Figure 1.4. Once thought to be a dawn horse, the specimen from the Messel Pit turned out to belong to the genus *Hyrachyus*, an early relative of primitive horses and tapirs. In 1974 private collectors offered to sell the object to the Senckenberg Museum. Today, this discovery is in the possession of the Hessisches Landesmuseum in Darmstadt.

as a scientist, I was forbidden any type of access to it. At one point I declared to the sellers straightaway that the Senckenberg Museum could never raise such an amount, and I added, "when we can scrape up the money, we will invest it in our own diggings."

A year passed. Under the pressure of the landfill plans and with aid of the furious public, the Senckenberg finally received excavation approval from the minister of culture. The Senckenbergische Naturforschende Gesellschaft authorized 20,000 DM for our research. With it, we were able to conduct a two-month excavation at Messel in 1975.

Finally I stood in the Messel Pit and saw the target of my dreams (Figure 1.5). Though like the other excavation participants I did not stand much of the time. Mostly we crouched or sat on the ground (Figure 1.6). Always we worked with the hope of discovering something exciting at the next splitting of an oil shale block: light brown bones, honey-colored teeth, or even a complete skeleton.

Figure 1.5. The Messel Pit in 1975. A lake formed in the pit after mining was abandoned. In the background, weathered granodiorite and bleached Permian sediments border the 1,000- by 700-meter opencast mining pit.

Figure 1.6. Members of the Frankfurt Forschungsinstitut Senckenberg excavation at the Messel Pit in the summer of 1976. The dark oil shale is combed through with excavation knives. The theolodite visible in the background is used to measure the discoveries. The red plastic chest contains the excavation book and tools for excavating, as well as newspapers and plastic bags for packing material. The white plastic tub to the far left in the foreground is used to store small discoveries; water in the canister next to it is used to keep the discoveries wet. In the aluminum suitcase is a camera.

Figure 1.7. The first primitive horse discovered by the Forschungsinstitut Senckenberg team in the Messel Pit was found on June 26, 1975. The block with the fossil was later covered with damp newspaper to protect it from dehydration. Manfred Grasshoff points to the priceless find.

It happened on a sunny Thursday, on June 26, 1975. We were only a small team on this particular day: My technical assistant Christel Schumacher and I were joined by two zoology students and an additional helper, as well as Manfred Grasshoff, a colleague from the Forschungsinstitut whose research areas were actually corals and spiders. Manfred had joined our excavation for the same reason other institute colleagues had done so: because of a general interest in Messel possibilities. We had dispersed to separate excavation locations. Someone whistled to get everyone's attention. I looked across at the others a few hundred meters away from my excavation location. What was wrong there? People were excited. They gestured and danced around. Evidently something extraordinary had happened. I raced over. There it lay. A complete skeleton including teeth was exposed in a block of oil shale (Figure 1.7). We had found a dawn horse during our first excavation in Messel. It was hard to believe.

After the commotion died down, we carefully went to work. Nothing was to be lost, nothing was to be damaged. Cautiously the slabs with the marvelous finds were covered in wet newspaper in order to keep the strong, hydrated oil shale damp. The slab would have quickly dried out under the solar radiation, leaving the fossils in it crumpled in our hands. We tucked away the papered find in a large plastic bag, carefully sealed and labeled with the date and position of our discovery. Then we set off with shaky knees upward on the slope to the old miner's house that at the time was the base of our operations. From here we would be able to transport the discovery in one of the institute's vehicles to the lab in Frankfurt.

At the institute, the news spread with lightning speed. However, we had to stave off the initial eagerness to study the fossil. First the find had to be protected. That meant it had to be transferred from the fragile oil shale into an epoxy resin and allowed to harden (Figure 1.8). Afterward, the fossil was cautiously uncovered from the back—in this case the original top side—millimeter by millimeter with the help of a scalpel and a sandblaster (Figure 1.9).

Our discovery made history. The media pounced on it immediately. Newspaper articles with pictures went around the world, as did television reports. What was important for us was that the entire world would hear of the plans to turn the location of our discovery into a landfill. In our battle against these plans, we now had an ally. The discovery's impact on science would reveal itself much later. It was the first dawn horse in the world that still contained gut contents. And to top it off, the skeleton had exceptionally well-preserved contours of the soft body (Figure 1.10). Today, after detailed scientific

Figure 1.8. (*Left*) This technique allows brittle fossils to be transferred onto a plate of synthetic resin. First one side of the discovery, in this case an alligator, is cleaned, then infused with the synthetic liquid resin. After that side cures, the slab is turned and the fossil prepared from the backside. **Figure 1.9.** (*Right*) A scalpel and small sandblaster are used to clean a fossil under the microscope.

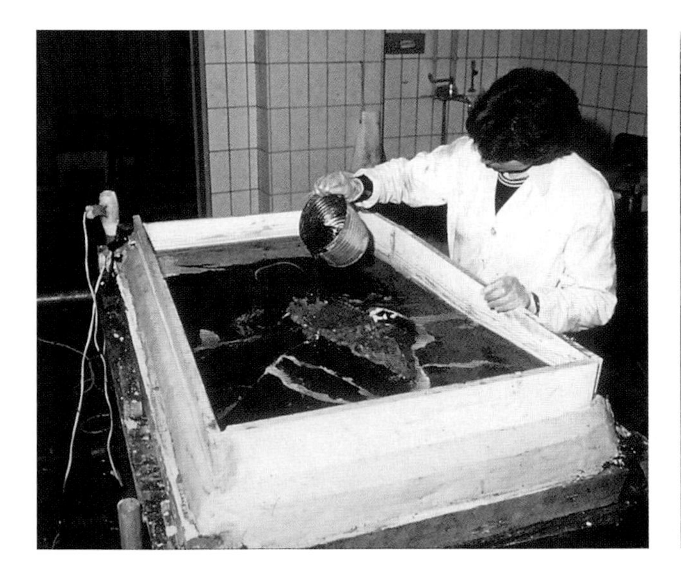

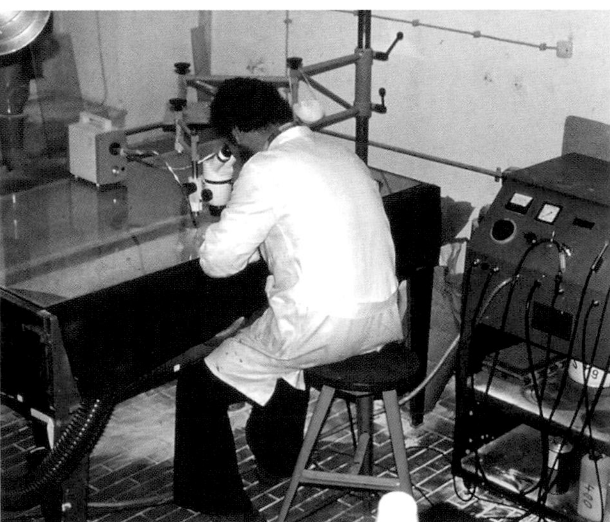

Figure 1.10. The first Senckenberg dawn horse discovered at the Messel Pit, a juvenile male of the species *Eurohippus messelensis*, after its preparation. The abdomen is still full of gut contents.

work, we know that it is a representative of an altogether new genus of horses. I have given it the name *Eurohippus*, European horse.

Stamp collectors all over the world became acquainted with the discovery in 1978 when the German post office issued a series of stamps with "our" dawn horse as the motif of the 2-DM stamp (Figure 1.11). This series also brought into the spotlight the importance of the discovery for science and the receptivity of a committed public. In 1995, after a long back-and-forth battle with changing governments and decisions, UNESCO designated the Messel Pit Fossil Site as a World Heritage location, the first, and up to now the only, monument of this rank in Germany (Figure 1.12).

For us, this meant time for unlimited excavations and with it, the prospect of discovering herds of dawn horses. Our excavation campaign of 1975 was extraordinarily encouraging: Besides the small dawn horse, we found a complete alligator, a wonderful bird with preserved feathers, and a bat—which would later became the motif of the 80-Pfennig stamp and was taken by astronauts on a trip around the globe. With these discoveries, our then director, Professor Schäfer, succeeded in persuading Volkwagen Foundation to accept fossils—after scientific analysis and publication—as cultural assets. Our excavation program at Messel was included in a

Figure 1.11. (*Left*) First-day edition (Ersttagsblatt no. 12/1978) of the stamp series Fossils issued by the Deutsche Bundespost (federal post office). The 80-Pfennig stamp (*top*) depicts a fossil bat (*Palaeochiropteryx tupaiodon*); the 2-DM stamp (*bottom*), the first Senckenberg dawn horse *Eurohippus messelensis* discovery from the Messel Pit. Both fossils were uncovered during the 1975 excavation. **Figure 1.12.** (*Right*) The certificate issued by UNESCO when the Messel Pit Fossil Site was declared a World Heritage Site.

foundation priority program for the rescue of cultural assets. From 1976 to about 1981, we were allocated around 1.2 million DM (about a half million U.S. dollars at the time) to invest in our Messel excavations. Later, the State of Hesse adopted the regular financing of the excavation and the excavation teams.

In spite of the success of our first excavation and those that followed, the Messel Pit is no El Dorado for fossils or a whole dawn horse herd, because the fossils are widely distributed both across the pit and among the strata. Also the excavations are anything but easy. In the summer, extreme heat spreads over the pit. When it is raining the air is muggy and the slates become slippery. And in the spring, at the beginning of the excavations, the first thing we have to do is to shovel away the snow caused by the winter cave-ins. These conditions, as well as the rarity of fossil finds of higher vertebrates, lead many of the excavation participants to grow tired and lose interest. Keeping participants' attention focused and their spirits up is essential. Insects, leaves, fish scales, and coprolites—fossilized feces—are more easily discovered when one is interested in and paying close attention to the work (Figure 1.13).

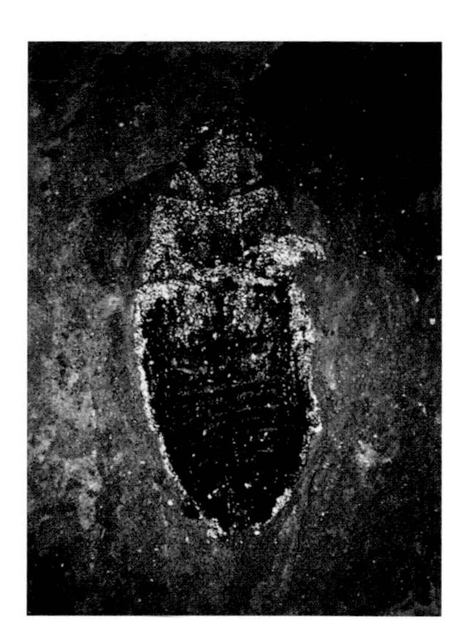

Figure 1.13. A preserved jeweled beetle (family Buprestidae) from the Senckenberg excavation in 1980. The natural coloring results from interference, emerging through reflection and superimposition of light waves.

In spite of all these adversities, the Forschungsinstitut Senckenberg has unearthed (up to now) no fewer than 16 mostly complete and articulated dawn horse skeletons from Messel. There are also the discoveries of our colleagues from other national and international institutes, as well as the numerous finds from private collectors. Since the last mining firm, Ytong, finally waived their rights to fossils from the property, more and more private collectors have been willing to show their treasures to the public. All together the number of specimens has totaled 60 more-or-less-complete skeletons. This is almost certainly an underestimate because private collectors made many of the finds during the transitional period between mining and institute excavations, and many of those finds remain hidden from the public. Naturally not only skeletons were found, although—quite unusually—they represent the majority of the discoveries. In addition there were isolated skulls, scapulas, pelvises, and other bones. The possibility at the beginning of excavation of finding an entire dawn horse has, in the meantime, become a reality. And there are still discoveries to be made. However, one should not be deceived by the numbers. Discoveries of complete dawn horse skeletons in the Messel Pit are extremely rare. One should not forget that the majority of the discoveries are the result of more than 30 years of excavations, with participants numbering more than 100 per year. In order to accurately assess this sensational fossil record of an early phase of horse evolution, one must know certain things. From the entire North American continent, there is only one dawn horse skeleton known that is of comparable age and exhibits a similar quality of preservation. To this day, no other continent has produced a single early Eocene horse fossil.

CHAPTER TWO

Introduction

WHAT ARE HORSES? This question can be answered in many different ways. Biologically and systematically speaking, horses are mammals, and are thus distinguished from other vertebrate classes, such as fish, amphibians, reptiles, and birds (Figure 2.1). Like other mammals, horses give birth to live young that are nursed by their mothers, in this case, the mares, during the first months of life.

As is typical of mammals, horses have differentiated teeth subdivided into incisors, canines, premolars, and molars (Figure 2.2). The first set of teeth that develop are called milk teeth, or deciduous teeth. The last three molars appear only once in the course of a lifetime, whereas all other teeth except for the wolf tooth are replaced once. Again, like other mammals, horses possess a relatively constant body temperature. In order to maintain this, a powerful metabolism is needed, based on large quantities of nutrition. An effective chewing mechanism is necessary for horses to acquire and process such food. A complex digestive system is also needed. This system is divided into a stomach, a large intestine, a small intestine, a cecum, and a colon. The cecum of horses is of particular importance because the digestion of cellulose, which is a significant component of a vegetable diet, especially grass, is enabled there with the help of bacteria. Therefore, the cecum of horses takes up a large part of the abdomen (Figure 2.3). Finally, horses have enough hair to effectively shield their bodies and assist them in maintaining their body temperature.

Within the class Mammalia, horses belong to the hoofed mammals (Ungulata)—a classification that is easy to see. Within the

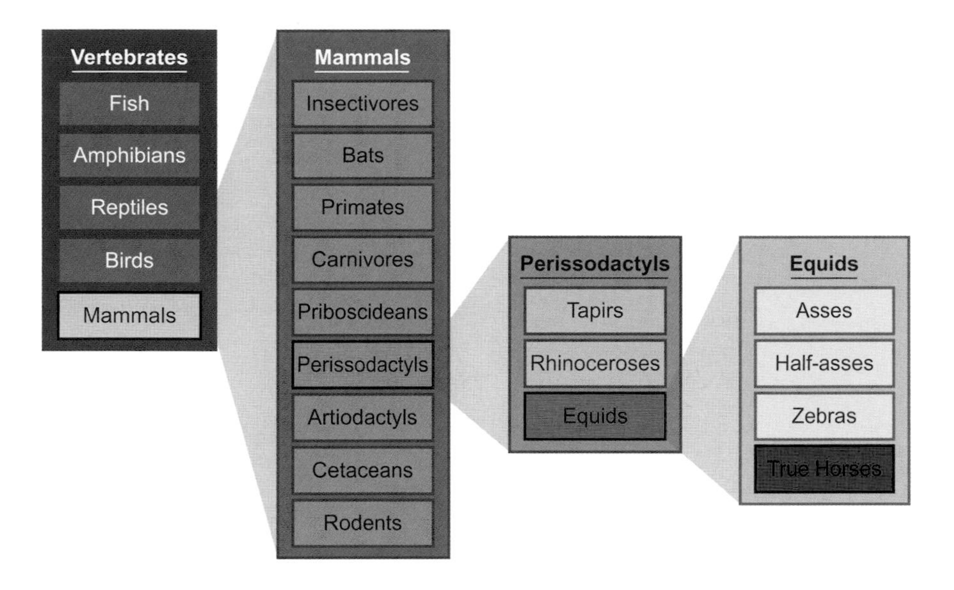

Figure 2.1. The systematic position of horses within the equids (Equidae), odd-toed ungulates (Perissodactyla), mammals (Mammalia), and vertebrates (Vertebrata).

Figure 2.2. A lateral view of the skull of a six-year-old horse showing the roots of the teeth: 1–3 = anterior cheek teeth (premolars); 4–6 = posterior cheek teeth (molars); 7 = wolf tooth (persisting first upper deciduous molar); 8 = hook teeth (canines or eye-teeth; usually occur only in the male); 9 = incisors. All teeth, with the exception of the posterior cheek teeth (molars) and the wolf tooth, possess a precursor in the deciduous dentition: deciduous incisors, deciduous canines, and deciduous molars. The complex of the frontal teeth—formed of incisors and in the case of males, also canines—is separated from the chewing apparatus—formed of the anterior and posterior cheek teeth—through a wide space called a diastema (from the Greek *diastema* = space). In domesticated horses, this space is used for the snaffle bit.

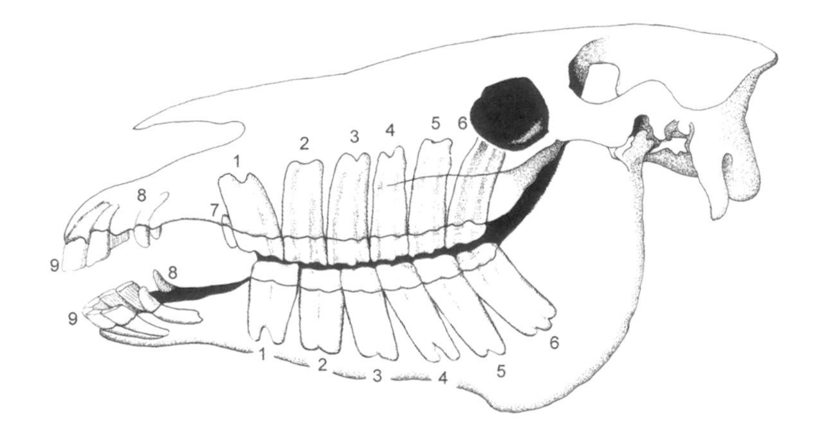

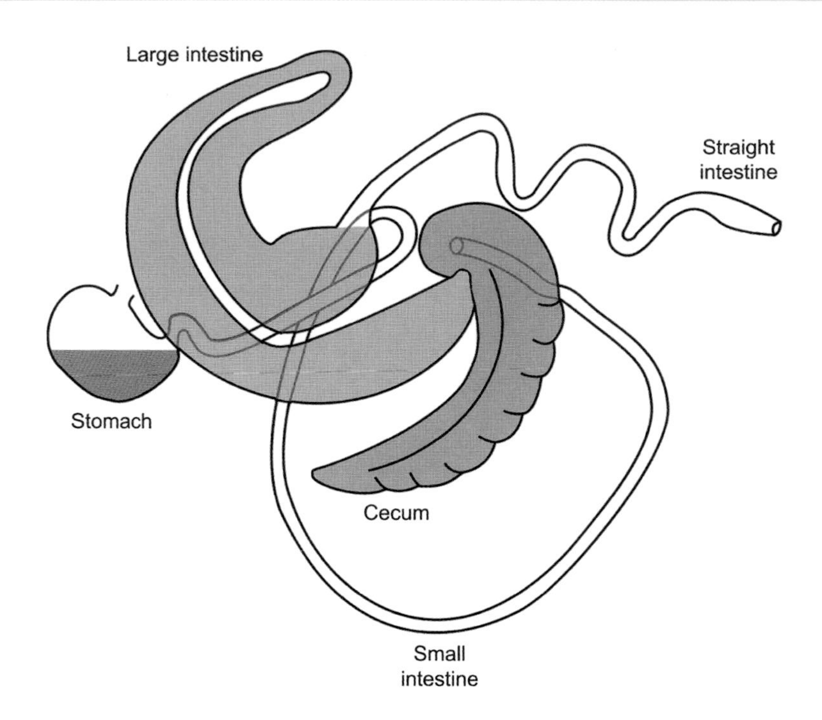

Figure 2.3. The gastrointestinal tract of a modern horse, showing stomach, small intestine, cecum, large intestine, and straight intestine (colon). The cecum provides a fermentation chamber for the digestion of cellulose by bacteria. Like all odd-toed ungulates (Perissodactyla), horses are cecum-fermenters.

modern living hoofed mammals, equids (Equidae)—together with rhinos and tapirs (see Chapter 9)—typify the odd-toed ungulates (Perissodactyla). Characteristically for these animals, the third or middle ray of the foot skeleton (metapodia and phalanges) is especially reinforced, whereas the laterals tend to become reduced. The first digit, that is, the thumb or the big toe, is always absent in perissodactyls (Figure 2.4). By contrast, in the even-toed ungulates (Artiodactyla) the third and the fourth rays are reinforced together. This may result in a complete fusion of the third and fourth metapodia, called the cannon bone. This is the case in llamas, camels, giraffes, antelopes, deer, sheep, and cattle. Pigs and hippopotamuses still display these bones reinforced but unfused. Like the perissodactyls, the artiodactyls are characterized by the complete reduction of the first digit. In modern fauna, the even-toed ungulates (artiodactyls) are more numerous and diverse than the odd-toed ungulates (perissodactyls), which are represented only by horses, rhinos, and tapirs.

Within the modern perissodactyls, the horses are characterized by a further reduction of their foot skeleton. They are the only one-hoofed mammals now on earth (see Chapter 6). All the one-hoofed equids are members of the single genus *Equus*. Within this genus, horses are differentiated further into species, such as the domestic horse (*Equus caballus*; Figure 2.5a), the ass (*Equus asinus*; Figure 2.5b), half-ass (*Equus hemionus*; Figure 2.5c), and different species of zebras (*Equus zebra, E. burchelli, E. grevyi*; Figure 2.5d), and the recently extinct *E. quagga*.

But what are dawn horses? For some people, images of the Duelmen ponies (Figure 2.6a), Exmoor ponies (Figure 2.6b), Przewalski's horses (Figure 2.6c), tarpans (Figure 2.6d), or the wonderful horse murals in the ice age cave paintings of southern France, northern Spain, and the southern Urals (Figure 2.9) may come to mind. Przewalski's horses, discovered by Nikolaj Michailowitsch Przewalski—colonel of the czar—on an expedition in the sandy desert of Central Asia, certainly have a primitive appearance. They, along with the closely related steppe and forest tarpans of Europe, were compared to the horse figures in the ice age cave paintings from which they are distinctly distinguished (see Chapter 10). The modern zoo-kept herds of Przewalski's horses go back to only 12 founding animals, 2 of which were foals that were collected for the Berlin Zoo by the

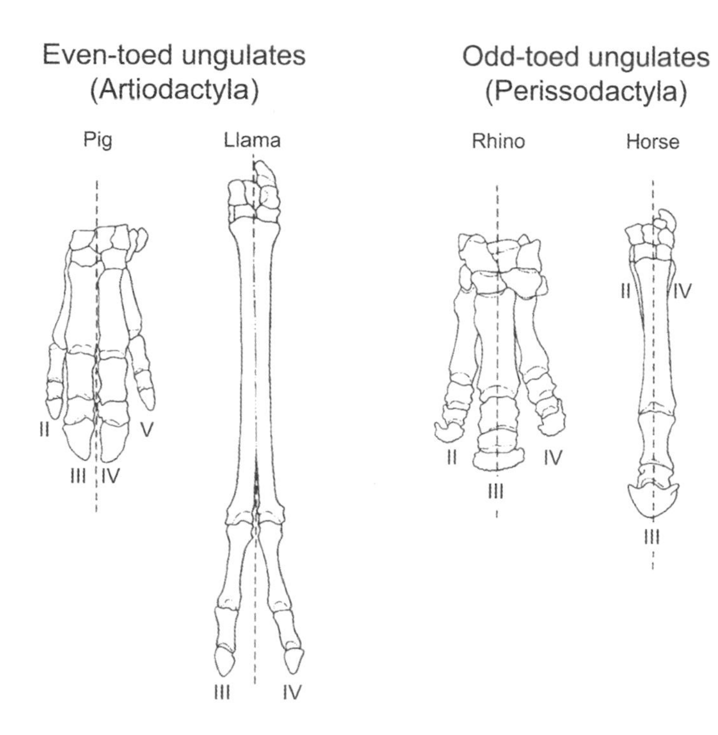

Figure 2.4. In contrast to the even-toed ungulates (Artiodactyla, *left*) (represented today by hippos, pigs, camels, deer, antelope, sheep, and cattle), in the odd-toed ungulates (Perissodactyla) like horses, rhinos, and tapirs, the middle or third ray of the original five-digit foot is reinforced. In the even-toed ungulates, the third and fourth digits are close together; in some cases, as in the llama, they form the completely fused cannon bone.

Figure 2.5. Modern horses: (a) domestic horse (*Equus caballus*) represented by a prize-winning Black Forest chestnut mare at the horse market at St. Märgen (Black Forest, Germany); (b) ass (*Equus asinus*), here a Poitou donkey in the Allwetterzoo in Münster (Westphalia, Germany); (c) Asiatic ass (*Equus hemionus*) in Hagenbecks Tierpark in Hamburg; (d) zebra, here steppe-zebras (*Equus burchelli*) in the Allwetterzoo in Münster.

Hamburg animal dealer Carl Hagenbeck on expeditions in the Kirghizian steppes in 1901 and 1903. The last wild living representatives were sighted in 1968 in the Gobi desert. Since then, wild Przewalski's horses have been classified as extinct. The tarpans are subdivided into the southern Russian steppe and the Polish forest tarpan. While the latter survived in Poland until the end of the eighteenth century, the last free-living steppe tarpans were hounded to death in 1876 by a hunting party. Only back-bred hybrids (Figure 2.6d), like those in the Munich Zoo Hellabrunn or in Poland, are left as a reminder of the last real European wild horses.

Admittedly, all these horses look rather archaic. Some of them are also quite old: The horse murals discovered at Grotte Chauvet (see Chapter 10) in the French central plateau are no less than 31,000 years old, an unimaginably long time in human history. However, all of these horses—except for the Przewalski's horse, which differs genetically and in the number of chromosomes (66 instead of 64)—are recognized as belonging to the same species as the domestic horse—*Equus caballus*. The real dawn horses are much, much older. Their fossilized remains have been found on all continents except Australia and Antarctica. Genuine dawn horses are those equids (members of the family Equidae) that do not belong to the species *Equus caballus*, are descended from the genus *Hyracotherium*, and lived prior to the end of the last ice age, more than 10,000 years ago.

Figure 2.6. Primitive-looking modern horses: (a) Duelmen pony in the Allwetterzoo in Münster; (b) Exmoor ponies in the Exmoor forest, Somerset, England; (c) Przewalski's horse in the Allwetterzoo in Münster; (d) Steppe-tarpan (back-bred hybrid) in the Sababurg Zoo (Lower Saxony, Germany).

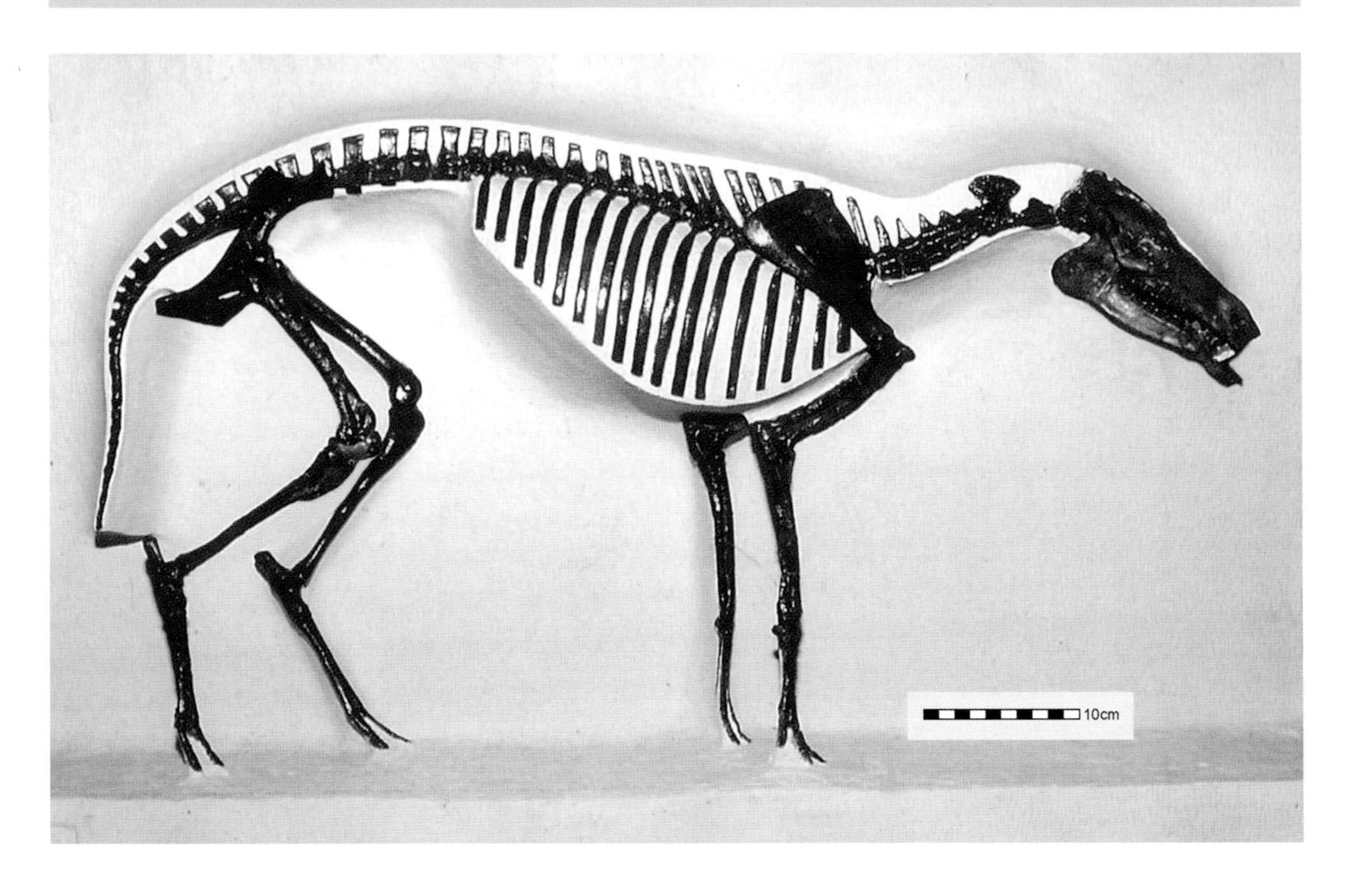

Figure 2.7. Skeletal reconstruction of a North American dawn horse of the genus *Hyracotherium*. Cast is in the Senckenberg Museum (Frankfurt am Main). The original is in the American Museum of Natural History in New York City.

The dawn horses of the stem group *Hyracotherium* lived around 52 to 55 million years ago. Because they are as small as a hare and therefore quite distinct from modern horses, the question arises of why these strange-looking animals are regarded as dawn horses at all (Figure 2.7). Not only is their size very different from modern horses, but the earliest representative of the genus *Hyracotherium* had 4 hooves on every forelimb and 3 on every hind limb. That is 14 hooves in total, in comparison to only 4 hooves on modern horses. Furthermore, their backs were not stretched out like they are in modern horses, but were rather heavily arched, similar to the duiker antelope or muntjacs of today (see Figure 6.12). Their very low cheek teeth with their comparatively simple bumpy occlusal pattern were also very different from modern horses (see Figures 6.1 and 6.17). Anyone today who saw a *Hyracotherium* with its 14 hooves and its diet of greens instead of hay would find it difficult to believe it was the stem father or stem mother of modern horses. How did this small animal arrive at its position as the ancestor of all of horses?

One could say that it attained this lofty position as the result of

160 years of scientific work. This answer is not false, but it is not complete. A more comprehensive explanation is that in the course of time, an interaction between theoretical expectations and concrete findings and discoveries led to the idea of *Hyracotherium* as the ancestor of all horses—an idea that is now accepted by almost all scientists. Even if this hypothesis is "right," which means it corresponds to the true course of events, it can never be proven because we don't have any eyewitnesses. There is only circumstantial evidence. (The historic scientific development of the investigation into the phylogeny of the horses is elucidated in Chapter 7.) Based on the present state of knowledge, *Hyracotherium* comes closest in appearance and age to the oldest dawn horses. This is because the older the rock strata from which they were excavated, the more the fossil horses resemble *Hyracotherium*.

Uncovering the evolution of the horse is not straightforward. Even with thousands and thousands of fossil horses from all over the world, the process of reconstructing the horse's evolutionary history is complex. Throughout all of time there were branch lines, or to put it another way, blind alleys of evolution. Occasionally the horse phylogenetic tree resembled a bush more than a tree, because many of the lines differentiated in parallel at the beginning of the Eocene in Europe or near the end of the Miocene in North America (see Figure 11.1).

THE QUESTION OF WHAT HORSES ARE can also be answered from a different perspective than the scientific view. For this, one has to reflect on the roles that horses have played and still play in human life. The human-horse connection is ancient. Although none of the dawn horses carried a rider, they served early on as prey of our ancestors, despite being quick and agile. This is corroborated by the numerous horse bones excavated from the dwelling places of fossil humans. Remnants of horses consumed by humans date from the middle ice age, around a half million years ago. Such evidence, however, is not satisfactory proof that we are dealing with prey. For such hypotheses, archeozoologists have to perform detailed analyses, such as recombining the individual bone flakes to reconstruct their fragmentation process. Also cut marks, left behind by the dismantling of carcasses with the help of stone tools, are identifiable using a scanning electron microscope (SEM). It requires almost forensic instinct to decide if a carcass was dissected by a person and whether or not the bone fragments are human food remains. And even if this can be confirmed,

Figure 2.8. The researcher Hartmut Thieme, in a coal mine near Schöningen, northeast of the Harz Mountains in 1995, looks at one of seven spruce wood spears that had just been unearthed. At almost 400,000 years old, these are the oldest spears in the world. This specimen (about 225 cm long) was found in such close contact with skeletal remains of wild horses that there is no doubt it was used for hunting these animals.

it is still not certain whether we are dealing with hunting or scavenging. Like in a crime thriller, the primary interest is to find out the cause of death. Couldn't early humans simply have eaten what predators or a steppe blaze left him? Were our ancestors, with their humble tools, capable of bringing down fast, agile, and shy animals, like horses? Clues to answer these questions were found only recently.

In the brown coal mine in Schöningen in the northeastern foothills of the Harz Mountains, the oldest wooden tools of mankind were discovered in 1995 during archaeological excavations directed by Hartmut Thieme from the Niedersächsisches Landesamt für Denkmalpflege. The tools are around 400,000 years old. Similarly old is a lance point of yew wood discovered at Clacton-on-Sea in Essex, England. Although lances could only be used at relatively close distances, no fewer than seven spears were found at Schöningen. The spears had been skillfully carved with stone implements from the heartwood of young firs. These spears turned up in such close proximity to the skeletal remains of horses that it appears evident that these horses were hunted and killed by early humans almost 400,000 years ago (Figure 2.8).

How could the early hunters unerringly hit the horses in midflight with their spears? The answer is again found by looking at the "crime scene." The discoveries were found at the edge of a formerly existing lake. Presumably, the animals were not on the run but stood drinking at the waterfront. Thus they were relatively easy prey. The hunters probably accomplished their goal by creeping up close to the animals so as not to reveal themselves and throwing their spears from a short distance away. Such opportune conditions for the horse hunt were not present everywhere. This became clear at other living places of early humans where hunting for other mammals has been shown, but no remnants of horses were present. The rather widespread hypothesis that our ancestors might have chased the scared horses with torches in hand into an abyss at advantageous locations—like the steep rocks near the village of Solutré, north of Lyon in France—and subsequently fed on the abundantly available meat of their victims has been refuted. There is a better explanation for the rich collection of horse bones at the foot of the rocks. Recent excavations have shown that the early hunters evidently took advantage of the existing bottleneck between the crag and the river by lying in wait for animals that passed through this location. The piles of teeth and bones were the result of the work of generations of hunters that accumulated at this place.

Even when horse meat was not abundant, it apparently was a significant source of food for early humans since the middle ice age. This probably explains why the horses, like other prey, have achieved a special mythological meaning throughout time. Everywhere portrayals of horses emerge among the cave paintings, sometimes together with strange symbols such as triangles, quadrangles, and dots, like in the Pech Merle cave in the French central plateau (Fig-

Figure 2.9. The frieze of the spotted horses in the cave of Pech Merle, Lot, in southern France. The illustration, some 25,000 years old, depicts one of the early horses. This early horse picture—as the distorted proportions show—is certainly not realistic.

Figure 2.10. Centaurs, figures of Greek mythology, were half horse and half man. This sculpture by the artist Carl August Sommer (1829–1921) is at the Centaurenbrunnen in Bremen.

ure 2.9), and sometimes on a protruding rock, like in the Chauvet grotto, in the "Panel of Horses" (see Figure 10.10). That horses still had a metaphysical meaning much later in time is shown by the centaurs, the horse-man beings of Greek mythology (Figure 2.10).

At the end of the ice age the relationship between humans and horses fundamentally changed. Horses were domesticated for the first time somewhere between 4,000 to 5,000 years ago. Fossil remains of wild horses from cave sediments in the French Jura Mountains around 5,000 years old are evidence of this. Human burials in horse-drawn carriages in the steppe parts southeast of the Urals date to approximately 4,000 years ago. It is highly probable that the domestication of horses did not proceed from a single population that lived at a definite time in a narrow outlined area. Genetic analyses suggest that domestication happened during a larger time interval and at multiple places over a large area. Geneticists have demonstrated that all modern living horses trace back to at least 77 different mares. As of 4,000 years ago, the evidence suggests that horses were domesticated in Greece, the Near East, and Egypt. Thus horses were domesticated much later than all of the other tamed hoofed mammals, such as cattle. The domestication of the horses at the end of the ice age resulted in shrunken herds, lifelong captivity, and submission, yet it also offered a new future for *Equus caballus*, even if without freedom.

For humans, the domestication of horses brought many advantages. Not only was the meat and milk of these animals available in almost infinite quantities at any time, but a much larger impact—as masterfully illustrated by North American Indians—was that horses became a means of transportation for our goods and household effects, as well as for people themselves (Figure 2.11). With this means of transportation, the mobility of our ancestors was enhanced to a previously unimaginable extent.

Figure 2.11. A group of Blackfoot Indians on horseback in the wilderness of Montana. At the end of the line, the two women are each pulling a travois, which consisted of two wooden tent poles bound with a pelt. The travois was used for transporting household goods.

Figure 2.12. The conquest of England by the cavalry of William the Conqueror at the Battle of Hastings in 1066.

Of considerable importance was also the impact horses had on the settlement of humans in central Europe approximately 5,000 years ago, and in the Mediterranean region even as early as 10,000 years ago. This impact was connected with livestock breeding and the cultivation of land. Over time, the horse relieved humans from plow pulling. Furthermore horses could have been efficiently applied to warlike conflicts in defense of settlements and also for the purpose of conquest (Figure 2.12). The employment of horses at least doubled the range of patrols. In other words, with horses it became possible to control a territory four times as large and to patrol from

Figure 2.13. A model of a stagecoach (1873) with driver in the Deutsches Pferdemuseum at Verden on the Aller. The original is in the Deutsches Postmuseum in Frankfurt am Main.

Figure 2.14. King Ernst August of Hannover on parade in the city of Hannover.

an elevated observatory, which meant a considerable strategic advantage. From that point, the development of cavalry was not far off.

As soon as the wheel was invented, the horse was used to pull combat vehicles and farm trailers and passenger coaches; eventually the horse tramway, forerunner of the cable car, came into use. The mail stagecoach was *the* telecommunications tool and means of transportation in earlier centuries (Figure 2.13). Without horses the German poet Goethe's travel to Italy would have been barely possible, just like countless other journeys. At a minimum, travel using a team of oxen would have made the trip considerably slower. Only with the beginning of the industrial revolution in the nineteenth century did horses cede all these roles bit by bit. Engines replaced horses in the cable car system; the car was substituted for the carriage; the electric mine car replaced the mine horse; the pump took the place of the donkey at the waterhole; and the tractor pushed aside the horse pulling the plow and the farm trailer.

Horses eventually became show and sports objects. The role of the horse as a status symbol—originally reserved for the nobility—

gained importance in sovereign acts, like during official state visits, parades on high holidays (Figure 2.14), and police operations. Horses have now become a luxury and a status symbol for the wealthy. They also serve as circus animals for entertainment (Figure 2.15). The riding horse, of course, has become a comrade for people and through that also emotionally attached to us. Horse racing, show jumping,

Figure 2.15. A pony performs in the Swiss National Circus Knie.

Figure 2.16. Ludger Beerbaum, winner of the Gold Medal, on Classic Touch at the 1992 Olympic Games in Barcelona.

and dressage examinations, which began in Middle Ages riding tournaments, remain a vibrant part of the human-horse relationship. Yet nothing characterizes the close connection between horse and human better than the fact that horses are the only animals that participate in the Olympic games (Figure 2.16).

The Depths of Time

BEFORE DISCUSSING FOSSILS and the evolution of horses, it is helpful to review the dimensions of time of the earth's history as well as the different methods of dating. The earliest dawn horses are referred to as "horses of the morning cloud." They lived during the Eocene, a time interval that began about 55 million years ago and ended about 34 million years before the present. The name "Eocene" refers to Eos, the deity of the morning cloud of Grecian mythology, who touched the early morning sky with her rosy fingers, thus driving out her sister Selene, the deity of the night, in order to pave the way for her brother Helios, the sun god, to rule the day (Figure 3.1).

The name Eocene was chosen for the "geologic time scale" by the English geologist and paleontologist Charles Lyell (1797–1875) in 1832. Lyell pointed to the fact that it was at this time when the first modern snails and mussels appeared. This epoch was also when the first members of modern mammal orders came onto the scene. Among them were the first primates, members of the same order to which we humans belong, as well as the earliest artiodactyls (even-toed ungulates). Also making an appearance was a particular form of perissodactyl (odd-toed ungulates), the early horses (members of the family Equidae). Lyell tried to define a sequence of layers of different ages based on the relative content of extant Mediterranean mollusk genera (snails and mussels). He began with the oldest layer he observed, the Eocene, followed by the Oligocene (from Greek *oligos* = little, and *kainós* = new). Above those layers of rock were the Miocene (from the Greek word *meion* = less), which contained more recent genera of mollusks, and the Pliocene (from Greek *pleion* =

Figure 3.1. Eos, goddess of the dawn in Grecian mythology, is depicted in a horse-drawn chariot while her sister Selene, goddess of the night, leaves the sky. Detail from a 1621 painting by Guercino (Giovanni Francesco Barbieri).The original is in the Casino Ludovisi in Rome.

more), with even more recent mollusks. At the end of the sequence Lyell defined the Pleistocene (from Greek *pleiston* = most) as containing most of the mollusks actually still living in the Mediterranean Sea. Later on, the French zoologist Paul Gervais (1816–1879) defined the Holocene (from Greek *hólos* = complete) as the time after the Pleistocene or ice age. The Holocene is the time in which we are still living today, the time of the present. Another interval was added by the paleobotanist Wilhelm Philipp Schimper (1808–1880) at the beginning of the sequence, prior to the Eocene. This is the Paleocene (from Greek *palaios* = old). When combined (Paleocene to Holocene), these time periods are called the Cenozoic (from Greek *kainós* = new, and *zóa* = animals), which is the modern age of the earth's history (see Figure 4.1). Before the Cenozoic was the "medieval age of the earth," or Mesozoic (from Greek *mèsos* = middle, medium). The Mesozoic, the time when dinosaurs walked on the earth, ended about

65 million years ago when the dinosaurs, together with a large number of other animal groups, went extinct, perhaps due to the impact of a huge asteroid, perhaps as a consequence of gigantic volcanic eruptions taking place in the North Atlantic and northwest India (the Deccan Trap), or perhaps as a result of a combination of these scenarios. A worldwide retreat of the oceans may also have contributed to this global turnover, which is still a wide field of active research.

It is one thing to talk about millions of years of the earth's history, but it is very difficult to imagine how long these periods of time really were. In order to understand the dimensions of geologic times and the extent of organismic changes involved it is helpful to use yardsticks for comparison. One yardstick might be the beginning of the Christian calendar with the birth of Jesus during the time of the Roman emperor Caesar Augustus (63 BC until AD 14). This event

Figure 3.2. The Alps with Lake Starnberg (Bavaria, Germany) in the foreground. During the Eocene, the Alps did not yet exist as a mountain range.

certainly happened far back in the past. However, the time when the horses of the morning cloud lived was about 25,000 yardsticks earlier. This is hard to fathom. Therefore, we need other yardsticks.

The origin of the Alps seems to be better suited as a yardstick (Figure 3.2). Evidence suggests that at the time of the Eocene the Alps did not yet exist as a mountain range. At that time, the waves of the former Tethys Ocean, or in other words the fringes of the early Indian Ocean, were still breaking against the southern coast of Europe. In the Savoyan Alps south of Geneva is a site at a height of 2,480 meters, where the remains of palaeotheres were discovered. These palaeotheres are relatives of dawn horses. They lived during the late Eocene at about 35 million years BP (before the present). The layer in which their fossils occur is covered by limestone that was once on the bottom of the sea. This means that the layer with the fossils was lifted up from sea level to 2,480 meters within 35 million years. In other words, an average rate of lifting of only 0.07 millimeter per year was enough to produce such a change of earth's face. For comparison, a human hair is about 0.1 millimeters thick.

Just as with the mountains, organisms changed and evolved during the earth's history. For example, based on what we know today about comparative anatomy, constructional and functional morphology, fossil documentation, genetics, numerical dating, and the molecular clock, it is quite certain that at the time of the dawn horses our own phylogenetic ancestors looked somewhat like modern Madagascan lemurs (Figure 3.3). Countless subtle changes led to the development of humans from lemurlike prosimians in the course of about 50 million years. During that same time, modern horses developed from ancestors such as *Hyracotherium,* which resembled rabbits more closely than horses.

If we convert the age of the earth (about 4.55 billion years) to 365 days (i.e., just one year), we see that the dinosaurs were living until 7:00 P.M. on December 26, while the dawn horses appeared no earlier than a day later, at about 2:00 P.M. on December 27. The whole evolution of horses as it is known from the fossil record would have taken place during the last four and a half days of the year. Thus the "horses of the morning cloud" are quite young.

How do we arrive at such calculations? What are the methods used for dating the earth's history? For a long time geology (earth history) and paleontology (the history of life on earth) were based only on relative dating, in particular stratigraphy. The basic law of stratigraphy was established during the seventeenth century by the Danish naturalist and medical doctor Niels Stensen (1638–1686),

Figure 3.3. A lemur (*Lemur catta*) at the primate center of Duke University in Durham, North Carolina. At the time of the Eocene our own phylogenetic ancestors looked and lived pretty much like this prosimian.

who is also known under his latinized name Nicolas Steno. Steno's law says that the oldest layers are usually the lowermost ones. They are covered by layers that are subsequently younger. Naturally, the application of this law is not as simple as it sounds. For instance the process of orogenesis, the development of mountain ranges, may lead to an inversion of the sequence of layers, and even to their duplication. In addition, the sequence of layers in one area does not tell very much about the age of similar layers in other areas.

The solution to such problems was discovered almost two centuries later by the English land surveyor William Smith (1769–1839). When Smith was completing the first geologic map of England and Scotland, he recognized that the various layers could be distinguished by the petrified organisms contained in them. Certain fossils (today called index fossils) characterize layers of a certain age. This way, it became possible to compare layers from different regions with respect to their age or—as it is expressed nowadays—to correlate them. Unknowingly, William Smith had applied the theory of evolution, which would be published about 50 years later by his fellow countryman Charles Darwin.

Beginning with William Smith—who became known as William "Strata" Smith—it was possible to arrange layers, or strata, depending on their ages. Still unknown, however, was the amount of time involved. It was Georges Buffon (1707–1788)—the great eighteenth-century French naturalist—who for the first time tried to find out the age of the earth. To do so, he used a large sphere of iron that was heated white-hot and measured the time until the sphere cooled down. That measurement—the time it took the sphere to cool—he reasoned, represented the age of the globe, after making adjustments for the much larger diameter of the earth. In this way, Buffon ended up with an estimate of the age of the earth of about 150,000 years. However, his method was not only indirect but also depended on quite a number of unproven prerequisites. Is it really justified to correlate the earth structurally and materially with an iron sphere? Did the temperature at the origin of the earth really correspond to that of a white-hot heated iron sphere? Did the cooling proceed consistently or was it interrupted by phases of reheating? What was the temperature in the surroundings of the young, still hot and fluid, earth? All this was unknown at Buffon's time.

An accurate dating of the earth's history only became possible with the discovery of the regular decay of radioactive elements by the English physicist Lord Ernest Rutherford (1871–1937) and by the determination of their half-life periods. A half-life corresponds to the time during which half of a certain amount of a radioactive element decays. Each radioactive element has its own characteristic half-life. These are absolutely constant and cannot be influenced by any factors. Therefore, knowing the half-life offers the possibility of measuring age based on the relation between decayed and not-yet-decayed quantities of a radioactive element as it naturally occurs in rocks. Such elements are often found in the minerals of volcanic rocks. The half-life of different radioactive elements varies greatly. Therefore, the age of what you are trying to measure determines what element is used for a numerical dating. The uranium-lead and the rubidium-strontium series are used for very old time intervals between 10 million and 4.6 billion years; the potassium-argon series, for ages between 50,000 and 4.6 billion years; and the uranium-thorium series, for ages younger than 350,000 years.

A special method of calculating age is carbon-14 or ^{14}C dating. In contrast to the radiometric methods mentioned above, ^{14}C dating permits a direct dating of fossils. This method relies on the fact that living plants consume carbon dioxide (CO2) and also incorporate the radioactive carbon-14 isotope from the atmosphere. Carbon-14 develops constantly from cosmic radiation in the earth's atmosphere. Assuming that development and decay of radioactive ^{14}C is balanced, the partial amount of ^{14}C in the atmosphere is regarded as constant. Therefore, by way of photosynthesis, ^{14}C is being incorporated into plants at a constant rate, together with the normal carbon isotope ^{12}C, and from there by way of food chains into other organisms, including mammals. Death brings this steady consumption of ^{14}C to an end. From that moment on, the radiocarbon clock is ticking. This means from that point, the $^{12}C/^{14}C$ relation begins to change due to regular radioactive decay of ^{14}C. On the basis of the relationship between the amount of still-radioactive ^{14}C and nitrogen (^{14}N), which is the result of its decay, it is possible to calculate the time since the organism died. Unfortunately, this method is only useful for ages between 100 and 70,000 years because the half-life of ^{14}C is only 5,730 years. By 70,000 years all the ^{14}C is pretty much gone.

Using these radioactive methods, it has become feasible to date layers numerically wherever they intercalate with layers of volcanic rocks such as tuffs (volcanic ashes) or basalts (streams of lava or volcanic bombs cooled down). By way of numerous calibrations of this kind, the whole scale of the earth's history has been more or less precisely dated. And with each additional dating, as well as methodological improvements, the standard scale of the earth's history becomes more accurate and more precise.

A permanent factor of uncertainty remains the geologic background. Neither rocks nor grains of minerals are static. The combination of different kinds of minerals in one and the same rock may result in a mixed age that does not date the origin of the rock. Therefore, the dating of isolated mineral grains is preferred nowadays. Such dating methods, as for example the one crystal argon-argon method using laser and mass spectrometry, are now so refined as to be, seemingly, almost infallible. Consequently, data of such quality replace more and more the mixed ages that were achieved formerly. Uncertainties arise only when the geologic context remains unresolved. For example, (1) What circumstances led to the formation of the mineral grain measured? (2) What happened to it subsequently? (3) What is the context between the mineral grain dated with greatest certitude with respect to the rock unit whose age is going to be determined? These are the questions that need to be solved in each case when radiometric dating is applied.

A series of other methods based on various preconditions completes this discussion of radiometric dating. These methods involve damage of the investigated material caused by ionized (radioactive) radiation. This damage results in changes of physical properties. Some of them grow with increasing age. Thus they become suitable for dating purposes. Important in the context of the evolution of the horses is the fission track method. It utilizes the fact that radioactive decay leaves behind in crystals as well as in glass minute tracks or fissures. Their number increases with age due to natural radiation. The tracks are counted under the microscope, making it possible to date ages between 10,000 and 10 million years, which is an important period in the evolution of horses.

In contrast to radiometric methods, paleomagnetics is not a dating method in the proper sense. It takes advantage of the fact that the polarity of the magnetic field of the earth reversed irregularly from time to time, which means that the North Pole and the South Pole exchanged their magnetic positions. The reason for this change is not yet known. Such a reversal of the poles may also occur in future times. Perhaps it is already taking place, because the intensity of the earth's magnetism is actually decreasing from year to year. In any case, naturally magnetic mineral grains (e.g., magnetite or hematite) in stiffening lava or in undisturbed sedimentation at the bottom of seas or lakes adjust their position to correspond to the earth's magnetic field. So the orientation of the earth's magnetic field becomes "frozen" in rocks such as basalt, tuff, sandstone, or claystone. Consequently, the magnetic field, as it was at certain times and in certain places, can be measured in rock samples taken in measured orientation and being radiometrically dated. This way, the changes of earth's polarity are reconstructed for geologic sections. Altogether, the paleomagnetic periods changing with time are counted, named, and graphically depicted on a global scale. In Figure 3.4, N (black) represents normal phases, while R (white) represents phases of reversed orientation of the earth's magnetic field. A global standard serves as a reference for identifying and dating strata from all over the world. A prerequisite for such dating is, however, that the sections are long enough to comprise enough reversals for their identification. It is then possible to correlate sections and to transfer numerical datings from spot to spot.

Similar methods depend on sudden increases (peaks) of certain elements (e.g., iridium) or chemical compounds (e.g., carbon diox-

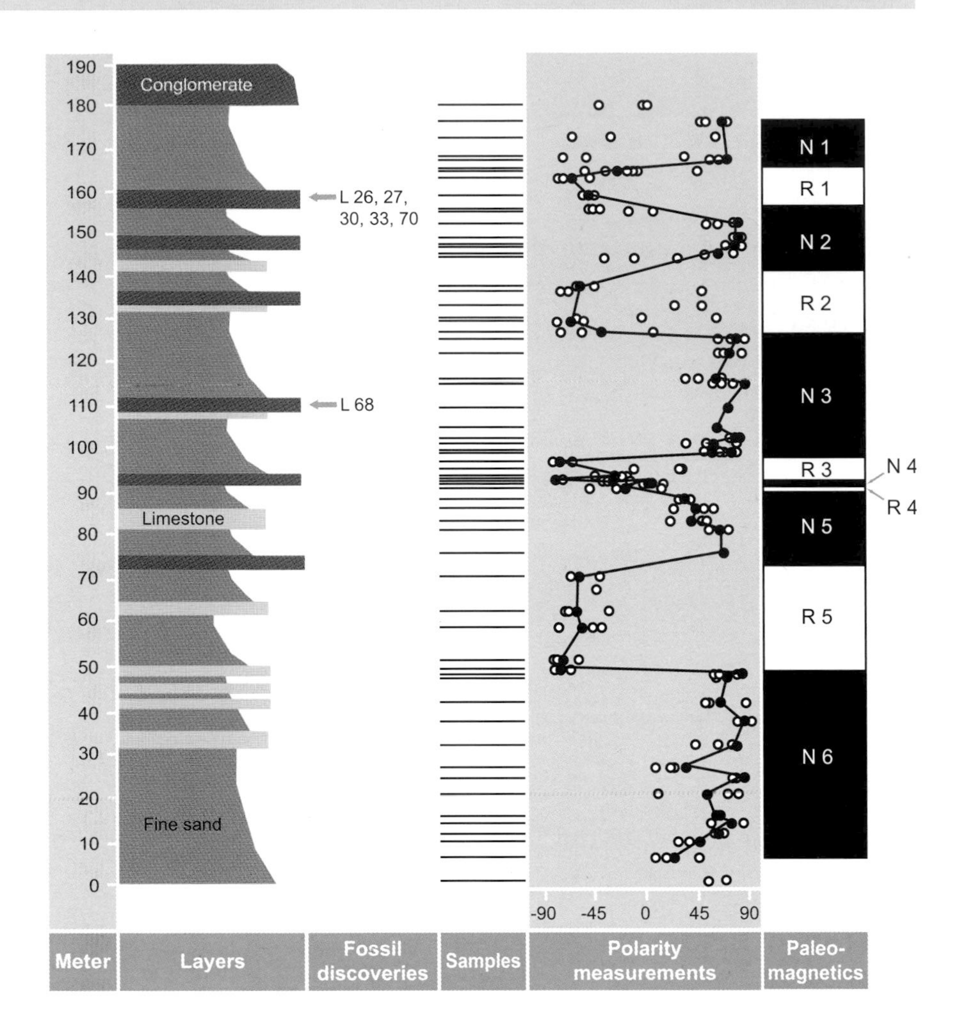

Figure 3.4. Paleomagnetics exemplified by a section of fossiliferous layers from the Late Miocene of Turkey. In the last column N = normal polarization, R = reversed.

ide). Such peaks may result from worldwide short-term events, such as impacts of asteroids or the degassing of methane from the bottom of the oceans. When measured and identified from geologic sections they permit intercontinental correlations and transfer of datings (Figure 3.5). With respect to the evolution of horses, such an event stratigraphy promises exact reconstructions of the time of appearance of certain genera of horses on different continents. This in turn would allow definite conclusions on the origin and migrations of horses at different times all over the globe (see Chapter 8).

Except for carbon-14 dating, all the methods mentioned above have the disadvantage of dating the age of certain fossils only indirectly. A molecular clock on the other hand determines the age of certain evolutionary events—such as the splitting off of new species—directly on the basis of the genome. This sounds fascinating and very convincing. It is, however, quite simple in detail and also employs a questionable method. What's the matter? Simply put, the molecular clock is nothing other than an estimate of similarity, as is the case with morphological taxonomy. The only difference is that we are dealing in this case not with morphological but with genetic similarity concerning the structure of the DNA (deoxyribonucleic acid). Provided that the rate of mutation is consistent—which means that the structure of the genetic apparatus is changing at a constant rate (which is not certain)—we can look at the degree of genetic similarity or difference and use it as a "genetic clock." The problem is, however, that the molecular clock has to be calibrated before it can be used. For this, paleontological estimates of the age of certain evolutionary events are determined—in the case of the horses, for instance, the time of the phylogenetic divergence into the three living species (i.e., horse, ass, and zebra). However this time is not yet exactly known. An age of 3 million years is generally assumed. As a result, the dating by the molecular clock cannot be more exact than

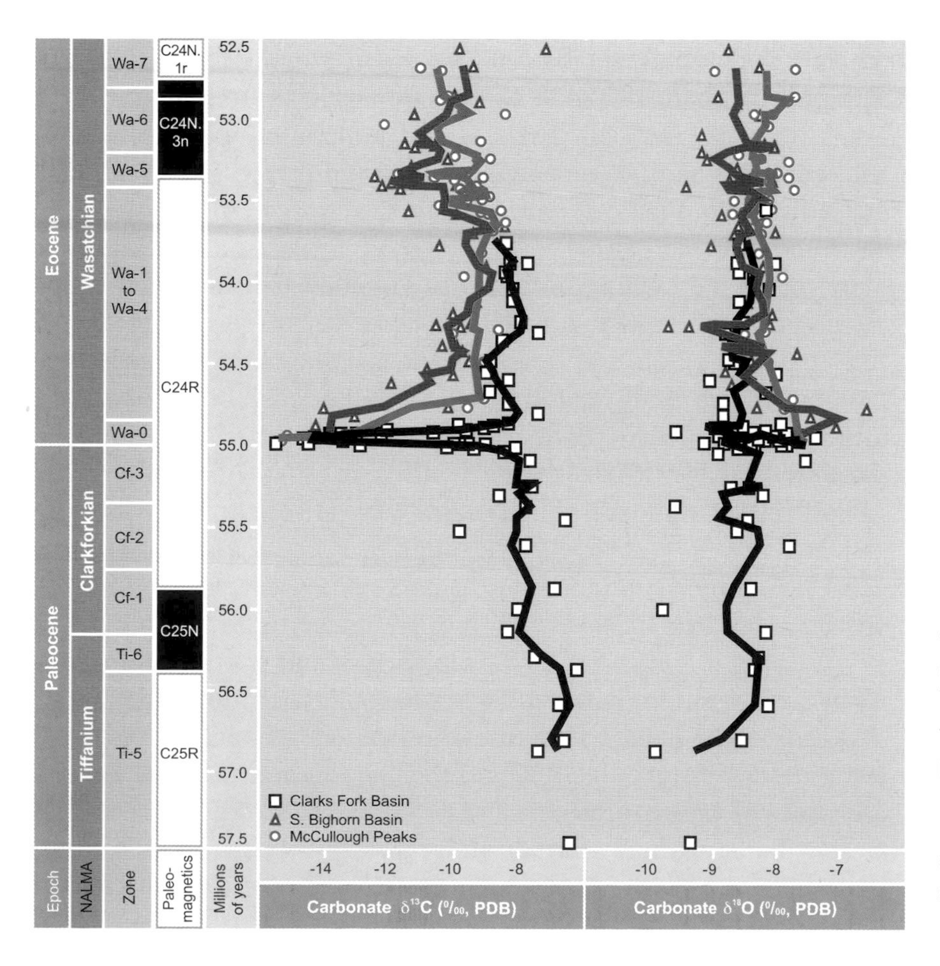

Figure 3.5. An example of event stratigraphy. The sudden and short-term increases (peaks) of ^{13}C as well as ^{18}O make it possible to correlate three sections of paleosols from Wyoming's Clarks Fork Basin, Bighorn Basin, and McCullough Peaks with each other, as well as worldwide with marine sections and the Paleocene-Eocene boundary.

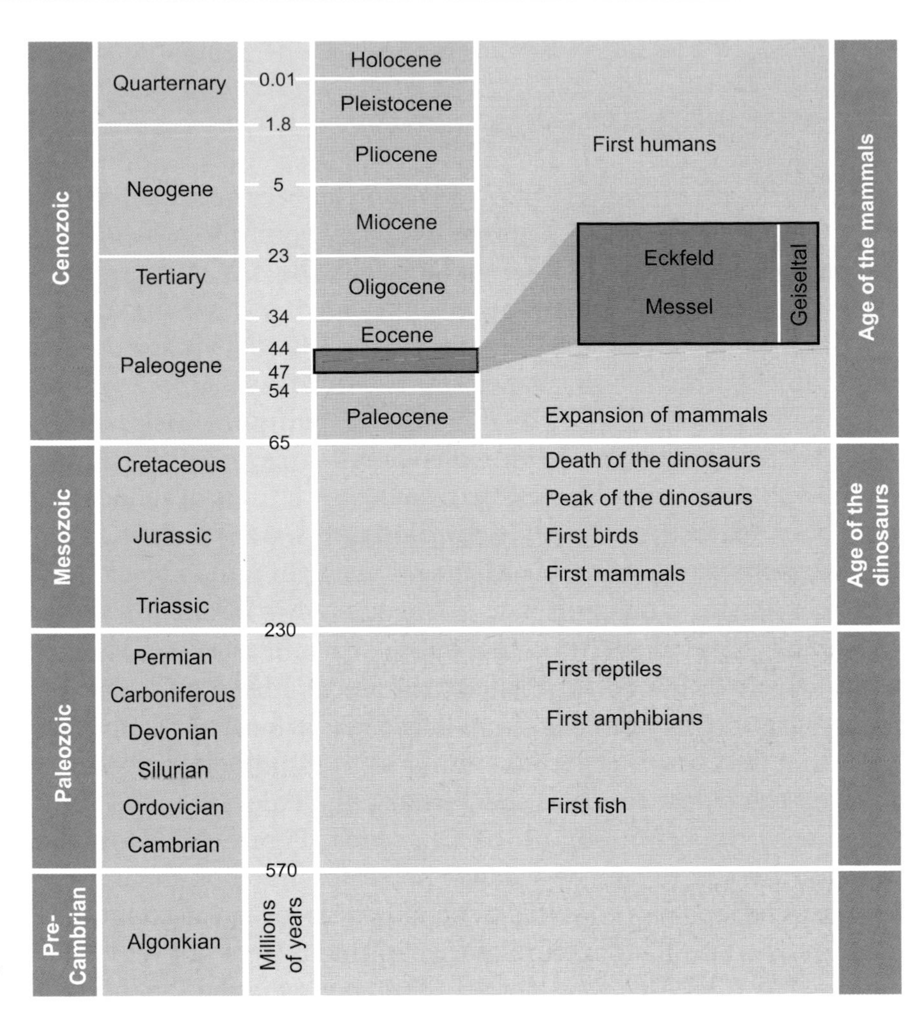

Figure 4.1. Geological timetable showing the position of the sites of dawn horses—Geiseltal, Messel, and Eckfeld.

connected to the continents only sporadically. Much like the Apennine Peninsula, a large portion of the Balkans were also under water. To the east, the Russian slab was likewise largely engulfed by the sea. It was separated from the Asian continent through the north-south directed Turgai Strait, which up until that time had connected the Arctic with the Indian Ocean.

In the west, the North Atlantic was still in an early phase of its development. At the beginning of the Eocene, Europe and North America were joined together by the Thule land bridge, which was located in the area of the modern-day Faeroe and Shetland Islands, as well as Iceland. Therefore, at the beginning of the Eocene around 50 percent of the mammalian genera were identical on both sides of the Atlantic, in Europe and North America. On the other hand, the land bridge functioned as an insurmountable barrier for the sea-dwelling organisms and drove them to separate development to the north and south. Thus the existence of the land bridge was supported by reciprocal evidence.

The sea was also influencing the development of the central European island. From the area of the modern-day English Channel in the west, it pushed forward into southern England and finally into the wide bowls of the Paris and Brussels basins. North of modern Belgium, the Netherlands and north Germany were also flooded during the Eocene. In the Paris Basin the sea formed a large lagoon, where the Paris limestone was deposited with innumerable mussels and snail shells. Many historic buildings in the French capital are actually constructed from this limestone. Later, toward the end of the Eocene, the famous gypsum of Montmartre formed during a period of desiccation in the Paris Basin. At the beginning of the nineteenth century, the French zoologist Georges Cuvier (1769–1832), a founder of mammalian paleontology, described one of the first and oldest mammalian fossil faunas of Europe from the Montmartre quarries.

When one considers that the Alps did not yet exist as a mountain range at this early time and that the Mediterranean stood in broad connection with the precursor of the Indian Ocean, it is understand-

Figure 4.2. Distribution of the sea (white) and land (gray) in Europe at the beginning of the middle Eocene, around 50 million years ago. The present-day geography is outlined in the background.

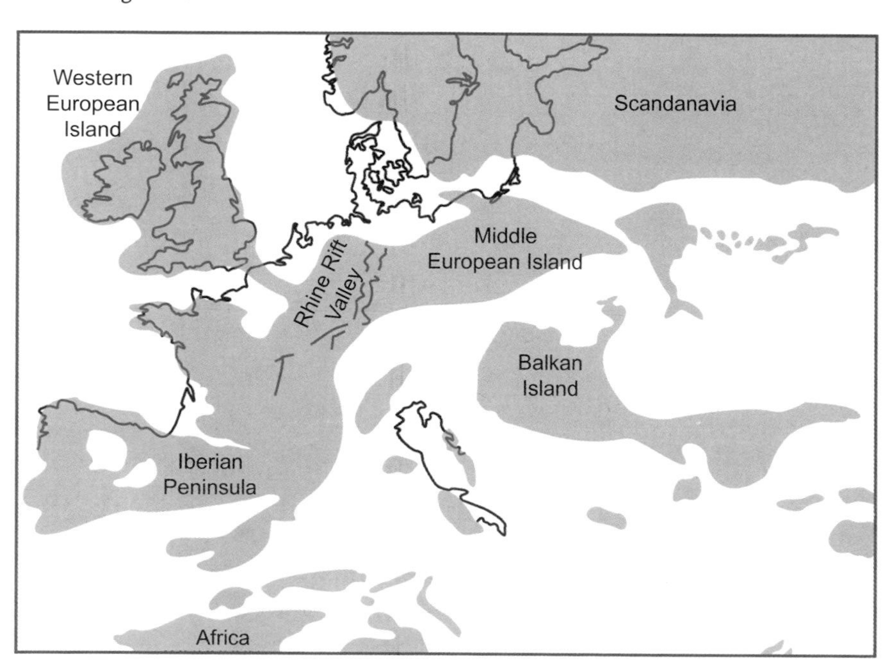

Toward the end of the Eocene the climate in Europe became drier and drier. Gypsum was not only deposited in the area of the Paris Basin but also in southern France, in the area of the modern-day central plateau, and in the Vaucluse. Corresponding to the changing ecological background, the composition of the mammalian fauna changed too. Concerning the equids, they were more and more superseded by the related palaeotheres (see Chapter 8).

Volcanic activity during the Eocene had a strong impact on Europe. Already at the beginning, the mighty ash layers of the Moler series, whose point of origin was presumably in the region of the Skagerrak, were deposited in northern Jutland (Denmark). Volcanoes rumbled, spit, and smoked not only in the surroundings of the Eocene Lake Messel but also in the whole area of the Rhine rift valley. As we now know, the Eocene Lake Messel resulted from a maar explosion that occurred around 47.8 ± 0.2 million years ago (Figure 4.6). (The deduction is based on analysis of a bore hole that was drilled in 2001 down to a depth of 433 meters.) In the southwest, Eifel, the oldest maar lake, developed around 44.3 ± 0.4 million years ago at Eckfeld—a few million years later than Messel (see Chapter 5). Overall, the Eocene was a time of transition between the quiet and smooth conditions at the end of the Mesozoic and the progressively turbulent Cenozoic, in which the geographical picture of Europe changed fundamentally.

Figure 4.6. Depiction of the creation of the Messel crater by a maar explosion 47.8 ± 0.2 million years ago.

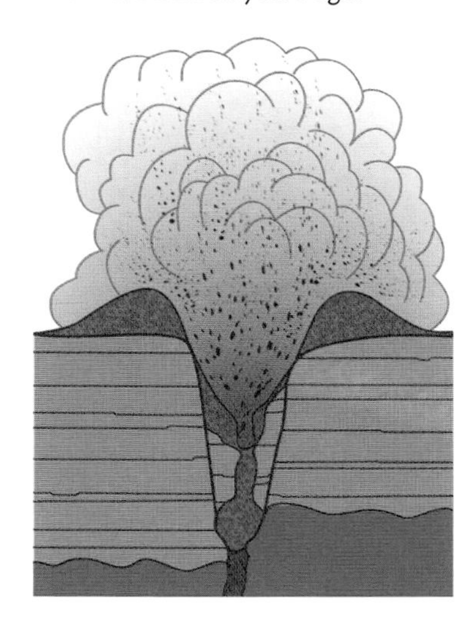

The transient nature of the Eocene is also expressed in the composition of flora and fauna. On one side, the pangolins (*Eomanis*) and anteaters (*Eurotamandua*) were still living in Europe as survivors of a fauna that had obviously already developed at the end of the Mesozoic. These species exist as "living fossils," in an almost unchanged form even today, although they are restricted to the tropics and subtropics. For other mammals, the Eocene was a time of virtually explosive evolution and expansion. Numerous modern orders appeared for the first time, such as the bats (Chiroptera), the primates, the proboscideans (Proboscidea), the carnivores (Carnivora), the odd-toed ungulates (Perissodactyla), the even-toed ungulates (Artiodactyla), and the rodents (Rodentia). The whales (Cetacea) also emerged at the beginning of the Eocene when their ancestors adapted more and more to an aquatic lifestyle, initially in freshwater and then in the sea. Obviously, the flourishing of mammals was enabled by the disappearance of the once-dominant dinosaurs, which went extinct at the end of the Mesozoic. The mammals rapidly developed and took over the vacant habitats on land, in the water, and in the air.

CHAPTER FIVE

The Dawn Horses of the Morning Cloud

In the heart of Germany three sites of Eocene plants, invertebrates, and vertebrates have been preserved. These are the former coal mines of the Geiseltal near Halle (Saale), the former oil shale mine of Messel near Darmstadt, and the Eckfeld maar in the southwest of the Eifel mountains, not far from Trier (Figure 5.1). All three sites permit a look far back into the Middle Eocene, and all have delivered extraordinary finds of dawn horses. There is no other continent, including North America, the homeland of the horses (see Chapter 8), that provides such a fossil record of truly exceptional quality.

The classic site of dawn horses in Germany is the Geiseltal (Figure 5.2), located about 20 kilometers southwest of Halle (Saale). During the Middle Eocene there was a lush, swampy rainforest with small lakes, ponds, and creeks. Today the Geisel valley is located there (Figure 5.3). The subsidence of the whole area as well as locally restricted subrosion of the underlying Muschelkalk formed wide basins. At the same time, diapirs were arising because of halokinesis. In the basins, a rich fauna and flora thrived and were preserved in coal layers up to the present. Between 1925 and 1939 the fossils were excavated by the Geologisch-Paläontologisches Institut of the Martin-Luther-University Halle-Wittenberg in cooperation with industrial coal mining. Through this cooperation, up until the beginning of World War II, the site Cecilie III, which is only one of about 30 occurrences of fossil vertebrates in the Geiseltal area, produced no fewer than 1,655 fishes, 250 urodeles (salamanders), 108 frogs, 91 lizards and snakes, 19 turtles, 27 crocodiles, 6 birds, 15

unique quality of preservation resulted in absolutely staggering information about the biology and soft body anatomy of animals that lived around 43 to 47 million years ago.

Most important with respect to the dawn horses was the discovery of a complete skeleton of a *Propalaeotherium* (Figure 5.4). But the recovery of countless isolated, reasonably well-preserved jaws and bones was also noteworthy. The skeleton became the symbol of the Geiseltal museum, while the bones and jaws made it possible to determine the relative amount of variation, which was truly remarkable (Figure 5.5). Moreover, the isolated bones revealed the articular facets that are normally invisible in articulated skeletons. Unraveling the construction of the articulations not only allowed a better distinction of the genera and species but also provided insight into the mode of locomotion.

Figure 5.3. The dawn horse *Propalaeotherium* browses in the foreground of this artist's rendering of a swampy forest in the Geiseltal. In the background are members of the even-toed genus *Anthracobunodon*.

In addition, the occlusal surfaces of the teeth of isolated jaws displayed details that were important for exact systematic determination. They also offered information about nutrition. Finally, in contrast to the localities of Messel and Eckfeld, the fossil record of the Geiseltal was not deposited in a rather short time interval. Instead, it contains at least six different stratigraphic horizons with fossils representing a total time interval of about 4 million years. Thus it is possible to reconstruct lineages of evolution that permit the classification of the biochronologic position of other fossiliferous sites and horizons, which can then be correlated with the standard time scale of European fossil lagerstätten. For this reason the Geiseltal site became the namesake of Geiseltalian, a time interval of the European Middle Eocene ranging from 43.5 to 47.5 million years BP. Altogether, the Geiseltal revealed no fewer than six different species of fossil horses, more than are known from any other European fossil lagerstätte.

Contrasting with the Geiseltal, the fossils found in the Messel Pit (Figure 5.6) come from a time span of "only" about 60,000 years (instead of the 4 million year span of Geiseltal). Geologically, this time span is almost no time at all. Evolutionary developments are hardly recognizable

Figure 5.4. The skeleton of the dawn horse *Propalaeotherium*. (The skeleton is the symbol of the Geiseltalmuseum at Halle.)

within such a short time. As fossil lagerstätte, the Messel site was discovered in 1875 during coal and iron prospecting in the Rotliegend (Permian) sediments. The first digs exposed a thin layer of lignite dark bituminous claystone. Unlike genuine coal, however, the claystone did not burn. Only by using smoldering kilns, based on furnaces developed in Scotland that were constructed especially for Messel, did the extraction of crude oil and kerosene by way of dry distillation become possible. Allegedly, during the Kaiser Reich the Messel Pit became, for a short time, one of the most important oil producers of the whole of Germany. At that time oil was used almost exclusively for kerosene lamps.

In the early 1970s, mining ceased at Messel after more than 100 years because profits were down and future prospects were dim. Consequently (and typically for that time), plans were made to transform the former mine into a giant garbage dump for the whole Rhine-Main area. After more than 22 years, in light of protests by civic action groups as well as scientists from all over the world, that absurd idea was finally abandoned. Not only were new methods of waste recovery developed but also a flood of extraordinary fossil discoveries made it clear just what was in danger of being lost forever.

body proportions. The largest species, *Propalaeotherium hassiacum*, has the longest skull of the four species, while its body is massive and stocky. The feet are not as slender as they are in *Eurohippus messelensis*, while *Hallensia matthesi* falls in between these two. This also holds true for *Propalaeotherium voigti*. Although there is no skeletal material for this species from Messel, a complete skeleton from Eckfeld shows that this species was almost the size of *Propalaeotherium hassiacum* but equipped with rather long and slender limbs. Evidently the smallest species, *Eurohippus messelensis*, was the fastest running and most agile of the Messel horses. *Propalaeotherium hassiacum*, with its relatively broad feet, was better suited for the swampy ground that was present at that time in the Geiseltal area. Perhaps this was the reason why the species was so common there. The locomotory apparatus of *Hallensia matthesi* appears to be comparatively unspecialized.

Altogether, 60 more or less complete skeletons of dawn horses have been recovered from the Messel Pit, more than anywhere else in the world. From the vast Asian continent there is not a single

Figure 5.8. Skeleton of a *Propalaeotherium hassiacum*, which was about the size of a German shepherd. This is the largest species of all the dawn horses from Messel. It is considerably more common than *Hallensia matthesi* but much rarer than *Eurohippus messelensis*. As is often the case with the Messel horses, the skeleton is disarticulated primarily in the area of the abdomen, apparently because of decomposition in this part of the body where most of the digestive system is situated. The tip of the mandible was completed by a cast.

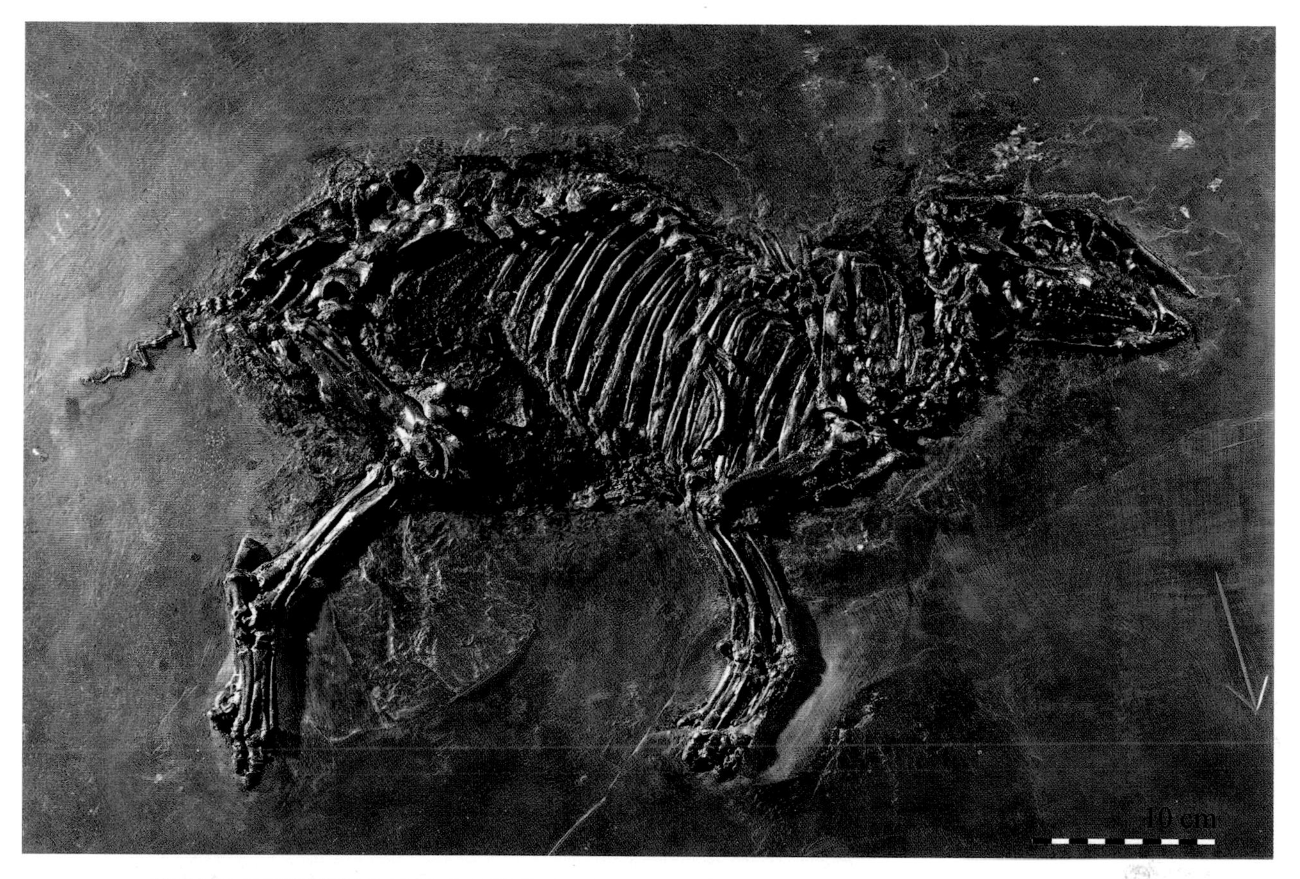

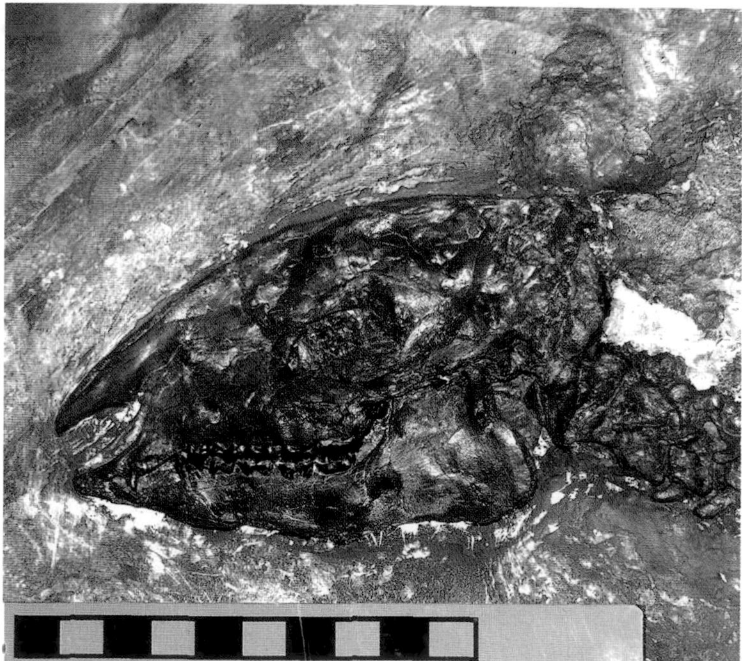

Figure 5.9. (*Above*) *Eurohippus messelensis*, which was the size of a fox terrier, is the most common and the smallest of the Messel dawn horses. It was no more than 35 centimeters at the shoulder. This find, from the 1982 Senckenberg excavations, is at the Senckenberg Museum (Frankfurt am Main).

Figure 5.10. (*Left*) Bacteriographic image of the outer ear in a specimen of *Eurohippus messelensis* found during the 1984 excavations.

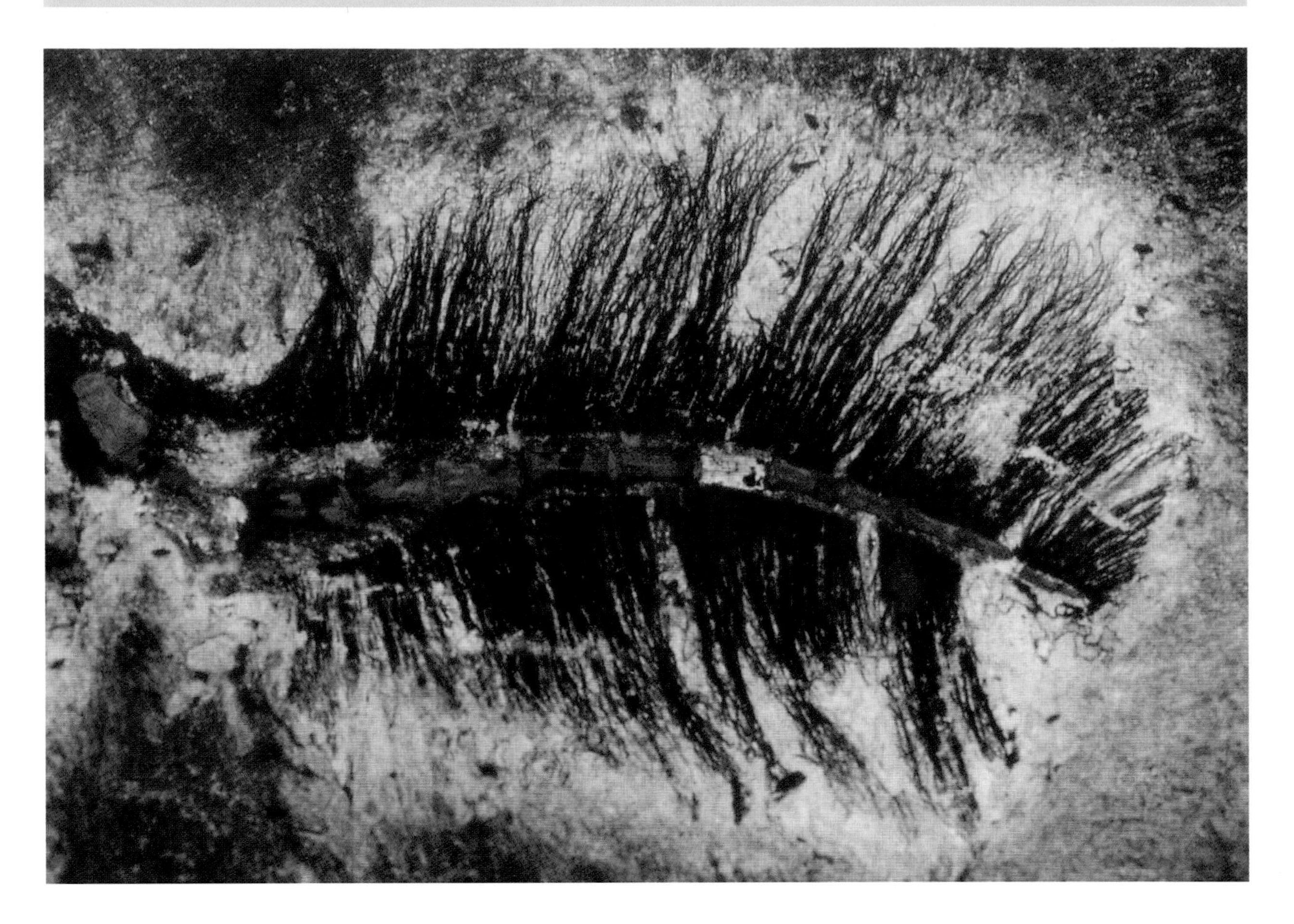

Figure 5.11. Bacteriographically preserved hairs from the tail of a pregnant mare (*Eurohippus messelensis*). The whole skeleton, found by a private collector in the Messel Pit, is now in the Museum Mensch und Natur in Munich (Bavaria).

skeleton of an Eocene horse known, and from North America there is only one of comparable quality (see Figure 8.11). The skeletons from Messel are extremely valuable for reconstructing not only what these dawn horses looked like, but also how they lived and how they moved. They also serve as the basis for identifying innumerable isolated bones from other areas, such as the Geiseltal. Once identified, these bones offer valuable information, such as details about articulations. Thus Messel and Geiseltal complement each other.

The significance of Messel rests on the information it offers about the paleobiology of very early dawn horses because of the quality of preservation of the finds. Most important is the so-called soft body preservation. Many of the specimens display the contours of their soft body as dark shadows or silhouettes, and in some cases even tips of the hair are present (Figure 5.11). Although isolated hairs were occasionally identified at Geiseltal and Eckfeld, at no other site is such detail available. At Messel it first became possible to recognize the form of the outer ears in two specimens of *Eurohippus mes-*

selensis. The ears were not as long as asses' ears nor were they cone shaped like in zebras. Instead, they closely resembled those of living horses in their proportions (Figure 5.10). The tails of Messel dawn horses were quite a surprise (Figure 5.11). A single specimen of *Eurohippus messelensis* as well as two individuals of *Propalaeotherium hassiacum* revealed that, in contrast to the long tail of modern horses, these dawn horse tails were rather short, with a puffy tassel near the end, somewhat similar to the tail of modern-day foals. One can only speculate about the functions of such a tail. Mosquitoes and other painful insects (swatted away by modern horses with the help of their long tails) must have crowded the rain forest of Messel. Perhaps the tail of these dawn horses was still primitive, not yet developed to such an extent as today. Or perhaps a short tail did not pose a problem in the undergrowth or a long tail was unnecessary for balance because the necks of these horses were short.

THE PROCESS THAT LED to soft body preservation at Messel is amazing. Michael Wuttke, who was working on his doctoral thesis at that time, succeeded in unlocking the secret. While looking at cells of the skin of frogs and bats under a scanning electron microscope (SEM), he didn't find features typical for skin but rather minute spheres and rods (Figure 5.12). Analyses with the Microsonde revealed that these corpuscles consisted of siderite, an iron carbonate (FeCO3). What was going on here? What processes had taken place? Morphology and the minuscule size of the corpuscles finally led to the interpretation that the strange structures were relics of bacteria. But why were these preserved in the form of siderite?

Michael Wuttke's explanation was as striking as it was convincing: As soon as the carcasses of these animals were deposited on the bottom of the lake, bacteria immediately began to decompose the bodies. The bacteria were producing carbon dioxide (CO2). This in turn was precipitating iron, which was amply available in the lake water due to decomposition of granites and Permian sediments in the surroundings. Thus siderite was formed, killing the bacteria and at the same time petrifying them. A thin layer of petrified bacteria developed at the point of contact between the decomposing bodies and the sediment. Subsequently, dark substances from

Figure 5.12. Petrified bacteria in the form of rods from the wing of a Messel Pit fossil bat (*Palaeochiropteryx tupaiodon*). The photo was taken with an SEM (scanning electron microscope).

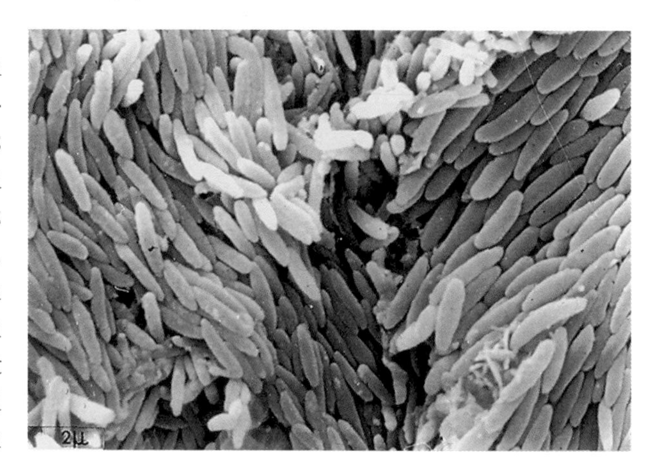

the mud—presumably kerosene of vegetable origin—impregnated and stained the porous layer black, copying the contours of the soft bodies with detailed precision. No soft body was completely preserved, but with the help of the bacteria an image was created by this chemobiological process. Analogous to similar photographic processes, it could be called bacteriography.

It was, however, not only the silhouettes of the animals' bodies that were preserved this way. In a find from colleagues of the Institut Royal des Sciences Naturelles de Belgique in Brussels, a dawn horse of the species *Hallensia matthesi*, even showed the contours of the cecum (Figure 5.7). This is far more than just a curiosity; the presence of a large cecum confirms that these dawn horses of Eocene times were already able to digest cellulose with the help of bacteria. Large amounts of gut contents in similar areas of the body suggest that the same holds true for the other dawn horses from Messel as well. Moreover, the quantity and the anatomical position of these gut contents confirm that we are dealing with a cecum and not a stomach (a mistake that was made at the beginning of those investigations). This arrangement can be compared to modern horses where the stomach is not only very small but is situated far in front, in the lower part of the thorax (Figure 5.13), far away from the area where the gut contents of the Messel horses are located.

The fossil record corroborates the hypothesis put forward by the American biologist Christine Janis in 1976. Based on the fact that living horses, tapirs, and rhinos are cecum fermenters, and that their phylogenetic development diverged before the Eocene, Janis argued that the common ancestor of all of these odd-toed ungulates must have been a cecum fermenter.

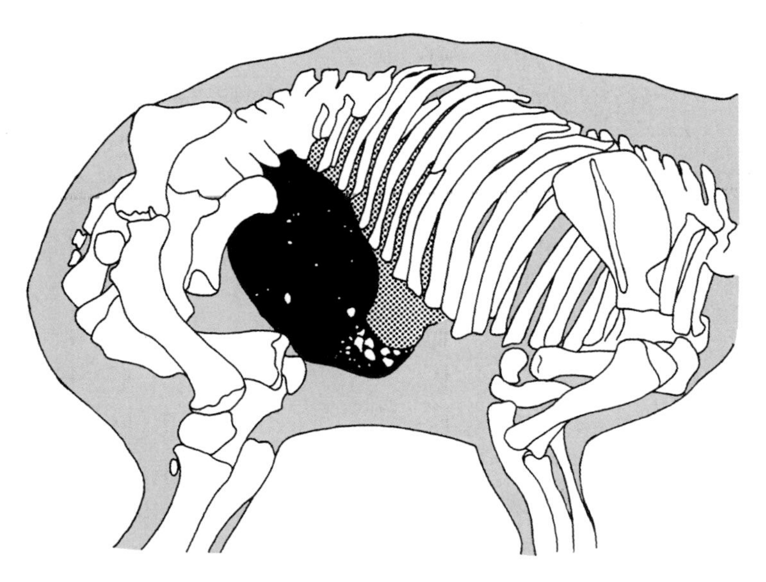

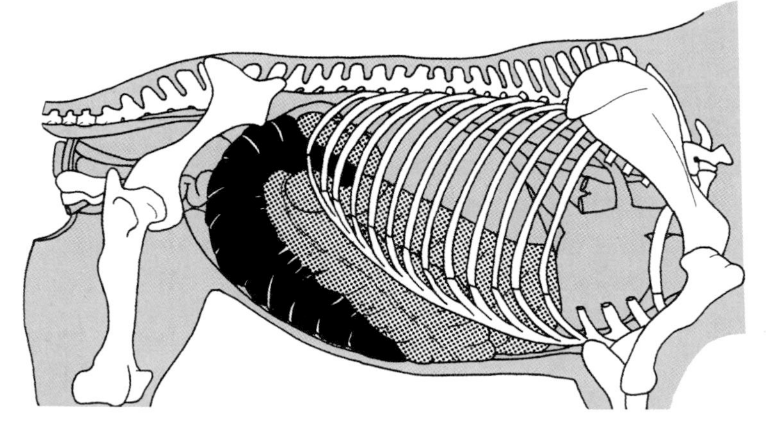

Figure 5.13. *Top*: Contours of the intestine of the dawn horse *Hallensia matthesi* from Messel (see also Figure 5.7). *Bottom*: The same anatomical area of a modern horse (*Equus caballus*). The cecum (black) is positioned behind the large intestine.

Figure 5.14. (*Left*) The remains of a laurel leaf from the gut contents of the first dawn horse (*Eurohippus messelensis*) found at the Messel Pit by the Senckenberg team in 1975 are seen under an optical microscope. The mosaic of the cell walls is recognizable to the experienced eye. (See also Figure 5.15.)

Figure 5.15. (*Right*) Fragment of a laurel leaf from the same fossil as in Figure 5.14. Using a scanning electron microscope (SEM), one can identify cell walls as well as a stoma (fissure).

The decisive prerequisite for this hypothesis was that this kind of digestion originated only once during the evolution of the perissodactyls. Nothing was known to contradict this assumption, and now the Messel fossils strongly support Janis' hypothesis. For an understanding of the evolution of the horses this is of special significance. It makes clear that the dawn horses were already able to digest grass rich in cellulose a long time before they became graminivorous. Graminivory (feeding on grass) occurred only as late as the Early Miocene of North America, about 20 million years ago. It becomes clear that the ability to digest cellulose was the decisive condition for the development of high-crowned (hypsodont) cheek teeth (see Chapter 6). Indeed, what would be the advantage of hypsodonty if the food were indigestible?

The Messel horses offer information not only about digestion during the Eocene but also about the kind of food itself. A dawn horse from Messel (see Figure 1.10) provided such information for the first time. When the fossil was prepared a large gray heap appeared that filled almost the whole abdomen. My first thought was that we were dealing with skin and hairs. Investigations with the SEM, as well as with the optical microscope, however, showed not skin or hairs but cell walls with fissures like those known only from leaves (Figures 5.14 and 5.15). What we were really dealing with was not the coat but the gut contents. This confirmed an assumption based on the morphological comparison of the cheek teeth with modern mammals, that these early dawn horses were browsers (see Chapters 6 and 7). Messel with its unique quality of preservation again delivered the evidence.

Later on, we discovered that there were different food sources available among the dawn horses because a few individuals of *Eurohippus messelensis* and *Propalaeotherium hassiacum* had seeds as well as leaves in their guts. In one case, seeds of grapes were identified in the gut (Figure 5.16a, b). The dawn horses of the morning cloud were evidently not only browsers (folivorous) but also fed casually on fruits (frugivorous).

Food remains, however, were not the only surprises we found in the abdomens of the dawn horses from Messel. We were already in the middle of our first digging campaign when one of our students came and asked me if the dawn horses could possibly be not only herbivorous but also occasionally carnivorous. Initially, I was dismissive. However, when one considers the cutting edges of the premolars and the carnivore-like canines, why should this be impossible? The student argued that a private collector's dawn horse discovery showed somewhat strange looking gut contents corroborating such an idea. We paid that man a visit, and he showed us his dawn horse skeleton. It was excellently preserved and very well prepared. There were some tiny teeth in the abdomen. Were these the teeth of a prey? I took a magnifying lens and looked at them more closely. Then, I realized that what I was looking at were the deciduous teeth typical for similar dawn horses. This was not the prey of a carnivorous dawn horse but the milk dentition of a fetus. We were looking at the skeleton of a pregnant mare.

Figure 5.16. The dawn horse *Eurohippus messelensis:* (a) skeleton from the 1980 excavation at Messel; (b) grape seeds from the intestinal area.

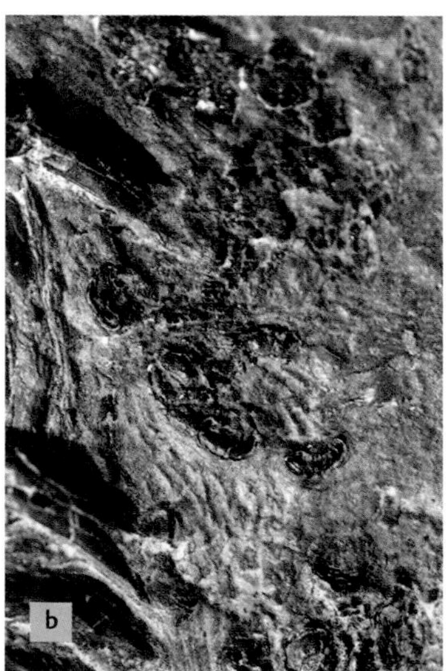

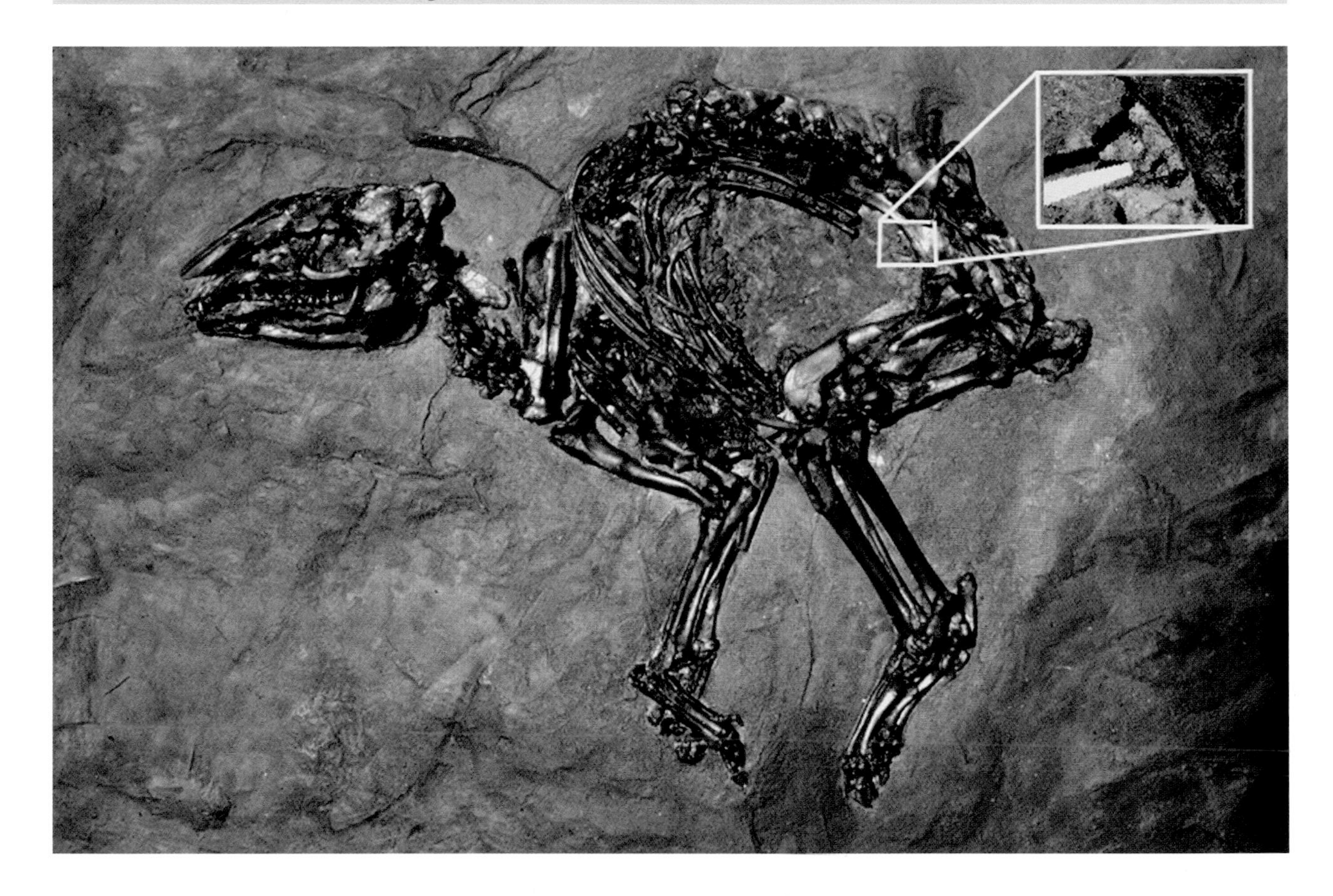

Figure 5.17. Skeleton of a pregnant mare of the species *Eurohippus messelensis*. The area of the intestine had been almost completely removed to probe into gut contents when germ stages of cheek teeth from a fetus (see inset) were discovered. The bones are very well preserved in this specimen, which was found during the 1980 Senckenberg excavation. The skeleton is so compressed from front to back that knees and elbows almost touch each other. This fossil is on exhibit at the Senckenberg Museum (Frankfurt am Main).

After this discovery, one pregnant mare after another was revealed. In the summer of 1980 we found a specimen of *Eurohippus messelensis* during our excavations (Figure 5.17). However its gut contents had been almost completely removed by a colleague before I saw the specimen. In the small part of the gut contents that remained I discovered the germ stages of a fetus' milk teeth. This was a mare in a comparatively early stage of pregnancy. All of that, however, was overshadowed by a find made by our colleagues from the Hessisches Landesmuseum Darmstadt in the summer of 1986. It was published in the following year by Wighart von Koenigswald, who was the curator at the Darmstadt museum at the time. In the abdomen of a complete skeleton, an articulated fetus was preserved in the position that is considered normal today: headfirst on the back.

Eight pregnant mares are currently known from Messel (Figures 5.18 and 5.19). All belong to the same species, *Eurohippus messelensis*, and all bear the remains of only one fetus. This suggests two things. Evidently, the dawn horses of the Eocene produced few offspring and invested substantial amounts of energy in parental care. Twins

ing toxins. Dawn horses drinking water from Eocene Lake Messel could have died seasonally by poisoning. The latitudinal position of Messel was about 40 degrees at that time so there must have been an annual climatic rhythm. However, restriction of sexual activity to a certain season and therefore a seasonal death can be refuted because the pregnant mares of *Eurohippus messelensis* were not fossilized in the same stage of individual development. Their fetuses and foals represent at least eight different ontogenetic stages within two years of individual age.

One other fact is remarkable for the pregnant mares of *Eurohippus messelensis:* some of these mares still display their complete milk dentition (Figure 5.20). Evidently, these females were not fully adult when they became pregnant. It was clearly important for each mare to give birth to as many foals as possible during her lifetime in order to transfer as many of her genes as possible to the next generation. If there was only one fetus per pregnancy it was desirable to maximize the fertile phase and to optimize it for propagation. Of course, the animals did not think of this in their sexual behavior but such a strategy was evidently advantageous.

In addition to pregnant mares and adult male dawn horses at Messel, there were very old and very young individuals. The age of each specimen is based on the ontogenetic development and attrition of the teeth. Foals are underrepresented for a natural population. Only four foals of *Propalaeotherium hassiacum* (Figure 5.21) and four of *Eurohippus messelensis* are known (Figure 5.22), although the latter species is two and a half times more frequent. Did the foals perhaps live at other, better-protected places in the surroundings of Eocene Lake Messel? Or was their death rate lower than that of adult animals? The latter does not appear very probable. It is also unknown why both skeletons of *Hallensia matthesi* are juveniles (Figure 5.23). Perhaps it's purely by chance.

If the sex of females can be determined on the evidence of fetuses, what can be said about the males? Are they also determinable? For a long time the assumption was that individuals with bigger canines were stallions, as is the case in modern horses. However, the exceptional quality of preservation allows other criteria to be considered. As in modern horses the construction of the pelvis is of special significance because the pelvic channel functions in females as a birth canal. For this reason, it is smooth and wide in modern-day mares and much narrower in stallions. Moreover, the processes on the inner side of the iliac blades additionally limit the pelvic canal in males (Figure 5.24). Exactly the same held true for the dawn horses

Figure 5.20. Cheek teeth of a pregnant *Eurohippus messelensis* mare from the Messel Pit. A complete deciduous dentition (*left*) is in front of the permanent molars, indicating that the mare was not yet an adult when it became pregnant. On exhibit at the museum Mensch und Natur of the Bayerische Staatssammlung für Naturkunde in Munich (Bavaria).

Figure 5.21. Skeleton of a juvenile stallion of *Propalaeotherium hassiacum* from the 1986 excavations by the Naturkundemuseum of Dortmund at the Messel Pit. Based on the stage of ontogenetic development of its dentition, the foal was 1 to 2 months old.

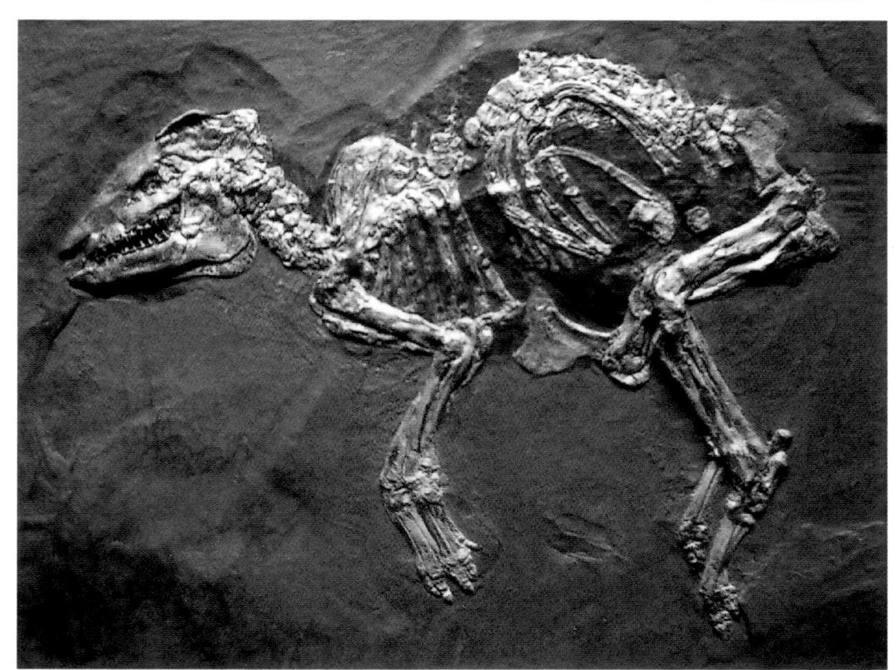

Figure 5.22. Foal of *Eurohippus messelensis* from the Messel Pit. Although less than 3 months old, it was already 2/3 the size of an adult. Find from the 1986 cooperative excavations by the museums of Karlsruhe and Stuttgart. On exhibit at the Staatliches Museum für Naturkunde Stuttgart.

Figure 5.27. A depiction of the movements of the dawn horses *Eurohippus messelensis*, which were the size of fox terriers.

Figure 5.28. Dawn horses of the species *Propalaeotherium hassiacum* were about the size of German shepherds.

for a maar, a deep lake with steep walls developed; it had a diameter of about one kilometer, more or less the size of the opencast mine of today. In the beginning, the lake was surrounded and somewhat isolated by the crater wall. Over the course of several hundred thousand years, the crater wall eroded and one or more small creeks flowed into the lake. As a result, during floods, carcasses of animals, such as dawn horses, may have drifted into the lake. Decomposing carcasses that filled with gases would have risen to the surface. Those that sank to a depth of more than 10 meters remained more or less complete. Experiments have shown that at such a depth the hydrostatic pressure of the water column becomes so strong that it prevents the carcass from blowing apart.

In shallower water, during decomposition gases would have developed under the humid and hot climatic conditions of that time, preventing the cadaver from sinking until it burst. Consequently, the bones would be scattered around on the bottom of the lake. This scattering has been observed in a few cases near the margin of the former lake (Figure 5.29). The distribution of more or less complete and articulated skeletons, therefore, occurs where the Eocene Lake Messel was deeper than 10 meters. At those depths the lack of oxygen prohibited the existence of scavengers. Their absence together with the lack of strong ground streams was responsible for the undisturbed fossilization at the bottom of the lake (Figure 5.30), which resulted in the unique quality of fossil preservation at Messel. What is normal for most fossil sites is abnormal at Messel and vice versa. For example, at Messel isolated bones and teeth are exceptional,

Figure 5.29. Bones of *Eurohippus messelensis* from the Senckenberg excavations in 1980. Apparently, in the paratropical climate gases in the decomposing carcass caused it to burst, scattering the bones on the bottom of the lake. The exhibit is on display at the Senckenberg Museum (Frankfurt am Main).

the maar explosion. Consequently, Eckfeld is about 3 million years younger than Messel. It corresponds with the age of localities such as Bouxwiller (France) or the obere Mittelkohle (oMK) of the Geiseltal.

The height of the land surface at the time of the maar explosion is estimated to be about 400 meters above sea level. At this height the ecological conditions must have differed from those at the Geiseltal, which, at that time, was situated at about sea level. Perhaps this explains why the flora of Eckfeld more closely resembles that of the paratropical rainforest of Messel than that of the swamps of the contemporaneous Geiseltal.

Concerning the dawn horses, Eckfeld clearly differs from Messel and the Geiseltal. The dominant species is not *Eurohippus messelensis/parvulus* (as at Messel) nor *Propalaeotherium hassiacum/isselanum* (as at the Geiseltal). At Eckfeld, *Propalaeotherium voigti* is the dominant species. In size, it is between the other two dawn horses. Its dentition resembles more closely that of *Propalaeotherium hassiacum/*

Figure 5.32. Excavation site in the Eckfeld maar.

isselanum, while the long and slender limbs correspond more closely with those of *Eurohippus messelensis/parvulus*. *Propalaeotherium voigti* is very rare at Messel, where it is represented only by a cranial fragment and a mandible that were found during mining times (indicating that they derived from comparatively higher and therefore younger layers). *P. voigti*'s rarity at Messel and its relative frequency in the untere Unterkohle (uUK) and the obere Mittelkohle (oMK) of the Geiseltal—the former corresponding in age with Messel and the latter with Eckfeld—make it highly probable that the differing ecologic background was responsible for the differing frequency of *Propalaeotherium voigti*. Presumably a difference in food availability among the sites was a critical factor. Unfortunately, detailed investigations of gut contents are still not far enough advanced to solve this interesting question. Food scarcity may explain why *Hallensia* is completely lacking at Eckfeld.

The highlight of all dawn horse finds at Eckfeld is the almost complete skeleton of *Propalaeotherium voigti* (Figure 5.33). The only known skeleton of this species that is more or less complete, it is exhibited in the Maarmuseum at Manderscheid (Eifel). One particularly interesting feature is a rather well-preserved fetus in the back of its abdomen (Figure 5.34). But this is not all. There is a partially preserved hard, fine-grained, bright gray layer that was clearly originally covering the whole fetus like an oval eggshell. The only plausible explanation is that we are dealing with a fossil relic of the placenta. Evidently it survived decomposition although some bones of the fetus were dislocated. Unfortunately, there are no histological details that identify it as a placenta. Bacteria are however preserved (Figure 5.35). Presumably, they played a role during fossilization.

The pregnant mare from Eckfeld is particularly interesting. That fetus has confirmed that *Propalaeotherium voigti* had a single, rather than multiple, offspring at a time. It is also possible to solve a systematic problem because it was still

Figure 5.33. Skeleton of *Propalaeotherium voigti*, a common species of dawn horses at the Eckfeld site. The specimen is a pregnant mare with a fetus that is fairly well preserved in its original position. (See also Figure 5.34.) The partially preserved placenta, which had enveloped the embryo, is unique among fossil finds. The original is on exhibit at the Maarmuseum at Manderscheid (Eifel).

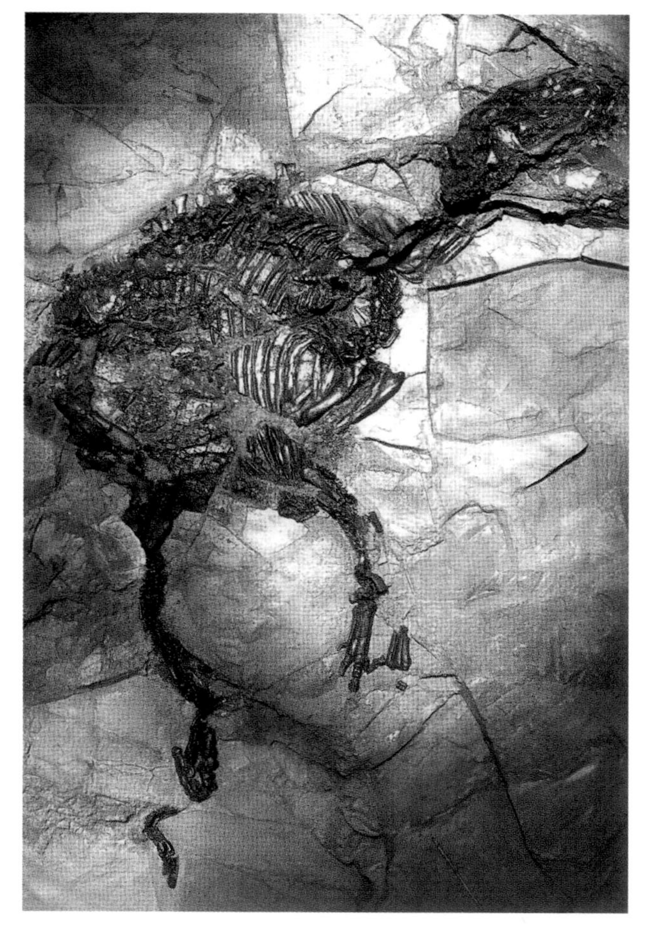

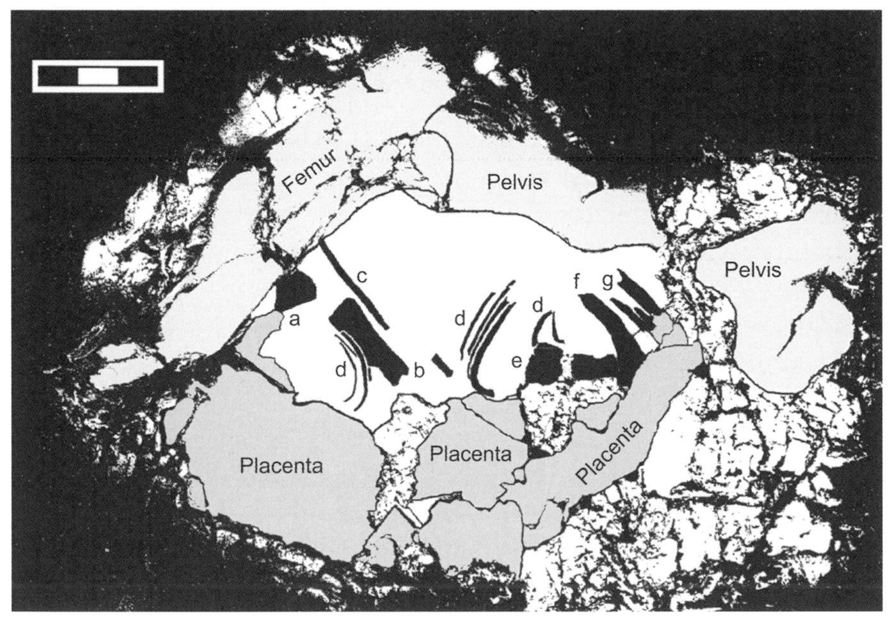

Figure 5.34. Fetus and placenta of the pregnant mare (*Propalaeotherium voigti* from Eckfeld) in Figure 5.33. *Top*: Specimen is covered with ammonium oxide to reveal details. *Bottom*: Bones of the fetus are *a* = shinbone, *b* = thigh, *c* = calf bone, *d* = ribs, *e* = shoulder blade, *f* = humerus, *g* = forearm; femur and pelvis of the mare are in light gray; the placenta is dark gray. The measure (*top left*) equals 3 centimeters.

Figure 5.35. SEM (scanning electron microscope) photo of petrified spheres of bacteria from the placenta of Figure 5.33, the pregnant mare of *Propalaeotherium voigti*.

vaguely plausible that *Propalaeotherium voigti* was not a species of its own but rather representing large males of *Eurohippus parvulus*. The discovery of a definite female resolves that question.

One other point makes the dawn horses from Eckfeld especially significant when compared with Messel or the Geiseltal: There are several isolated skulls of *Propalaeotherium voigti*, *Propalaeotherium isselanum*, and *Eurohippus parvulus* that are not compressed and flat, but are preserved three dimensionally and almost undeformed (Figure 5.36). They show the form of a long stretched wedge as is still typical of horses, and not the baggy profile indicated by the flattened skulls from Messel.

The fossil record of dawn horses from the Eocene of Europe is not restricted to the three fossil lagerstätten of the Geiseltal, Messel, and Eckfeld. Important sites are also found in France, Spain, En-

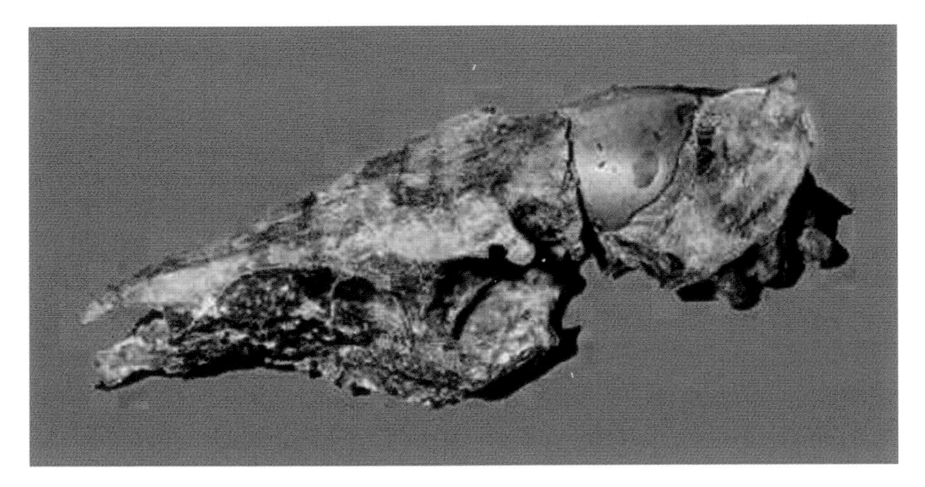

Figure 5.36. Skull (about 18 cm long) of a *Propalaeotherium voigti* from Eckfeld seen from above. Unlike the dawn horses from Messel and the Geiseltal, many of the crania from Eckfeld are relatively well preserved. The original is on exhibit at the Museum für Naturkunde at Mainz.

gland, and Switzerland, although those finds mostly consist of isolated dentitions, teeth, and bones. Only rarely is a skull preserved. Large numbers of finds come from karst fissure fillings, for example, at Egerkingen in Switzerland. The classic area for the fossil record of the evolution of horses during the Eocene of Europe is the London Basin with the skulls of *Hyracotherium leporinum* and *Pliolophus vulpiceps*. The Paris Basin is also very important. Dawn horses occur there, but even more abundant are their close relatives, the palaeotheres (see Chapter 9), which were studied by Georges Cuvier (1769–1832). All of these fossils are preserved in a sequence of layers in a large flat lagoon that contains fossils from the Paleocene/Eocene up to the Eocene/Oligocene boundary. Together these finds contribute clues to the puzzle of the evolution of the horses and their biotopes during the Eocene of Europe (see Chapter 8).

CHAPTER SIX

Constructions and Functions

IT IS CHARLES DARWIN (1809–1882), the famous nineteenth-century English naturalist, to whom we owe the breakthrough on the theory of evolution. Before Darwin, ideas about a natural development of organisms already existed, but those ideas were somewhat offtrack. For example, the French biologist Jean-Baptiste Lamarck (1744–1829) was convinced that organs were developed based on use or disuse. In his 1859 work *On the Origin of Species by Means of Natural Selection*, Darwin summarized all the countless observations and discoveries he had made while traveling around the world from 1831 to 1836 on board the HMS *Beagle*. These observations led to his theory on the origin and diversification of species, which many people still consider revolutionary. In Darwin's explanation, species originated as a consequence of variation and selection of certain biological characters. In this process the natural environment acted like a human breeder of animals. Only those organisms that were endowed with the best equipment succeeded in their fight for survival over time (survival of the fittest).

Strictly speaking, another English scientist preempted Darwin. The zoologist and botanist Alfred Russel Wallace (1823–1913) also traveled the world. His voyages took him to the Amazon River and Rio Negro from 1848 to 1852 and to the Southeast Asian archipelago from 1854 to 1862. He was the founder of zoogeography, the science of the differentiated distribution of animal species all over the globe. In the course of his investigations, Wallace recognized the origin and modification of new species by way of natural selection quite independently from Darwin. He published his ideas for the first time in 1855, four years earlier than Darwin, in an essay titled "On the Law Which Has Regulated the Introduction of New Species," followed three years later by the book *On the Tendency of Varieties to Depart Indefinitely from the Original Type*. But influential friends and Darwin's higher social status helped him, and not Wallace, to be recognized as the father of the theory of evolution. The driving force of evolution was for Darwin, as well as Wallace, the natural environment. Is it really possible, however, to fully explain the origin and change of species and their characters in that way? Not all features of an organism are directly affected by environmental change, and not every ecological change leads to new species. On the other hand, new species originate under the stable environmental conditions typical for tropical forests or global oceans. Additionally, Darwin's theory does not explain the incontrovertibility of evolution. Why do former characters and species not reappear in the course of cyclical changes of climate and vegetation, as was so characteristic of the ice age? Why is phylogeny not circular? Why does it not go backward if it depends on environmental change?

When one considers an organism, some features (e.g., the dentition, the end phalanges of the toes, the sense organs) are in direct contact with the environment. Consequently, they reflect environmental changes. Animals that live in caves or in the depths of oceans lose their eyesight. Some animals camouflage themselves by mimicry. Aquatic mammals, such as seals or whales, develop fins although their construction differs from that of fish. The same analogy applies to horses. For example, the phylogenetic development of their dentition seems to correspond to ecological development, particularly that of vegetation and food (Figure 6.1). But some investigations have showed a considerable time discrepancy between the first appearance of grasslands, on the one hand, and the development of high-crowned (hypsodont) cheek teeth with complex occlusal patterns, on the other. In any case, the incontrovertibility of evolutionary development is still evident. For example, dawn horses of the European genus *Hippotherium* returned about 12 million years ago to the forest-dwelling and browsing lifestyle of their ancestors. But the formerly low-crowned (brachyodont) and bump-shaped (bunodont) cheek teeth of Eocene horses do not reappear. Instead, the existing high-crowned type of cheek teeth with their typical flat occlusal surface inherited from their grass-eating ancestors is transferred into a shearing type suited for leaf eating (folivory) by the development of a secondary relief (Figure 6.2).

Innovations do not appear in an organism overnight. Their evo-

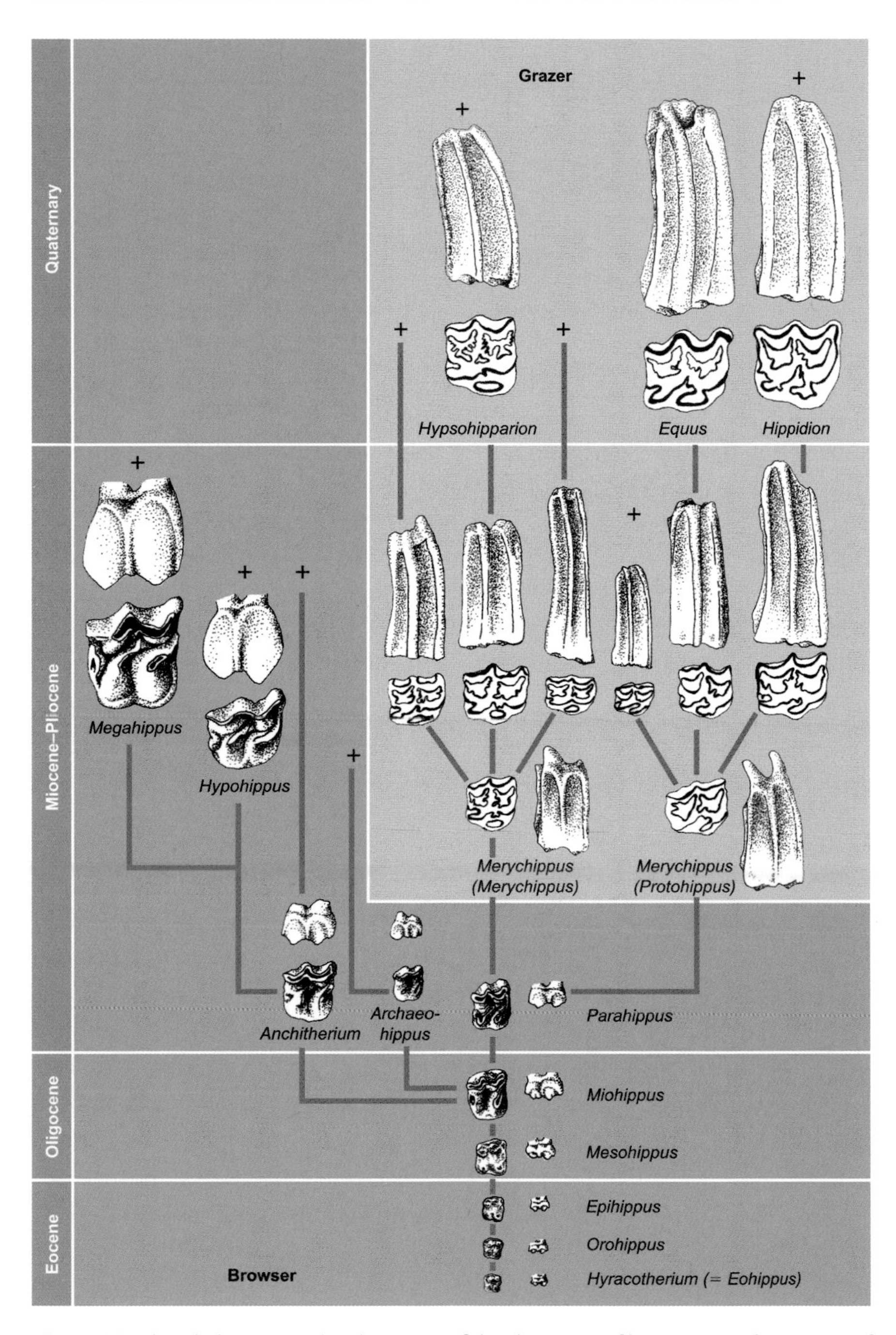

Figure 6.1. The phylogenetic development of the dentition of horses over the course of 55 million years. Each taxon is represented by one upper molar as seen from its external and occlusal surface. Note the almost explosive increase of crown height and enamel folding around 18 to 20 million years ago.

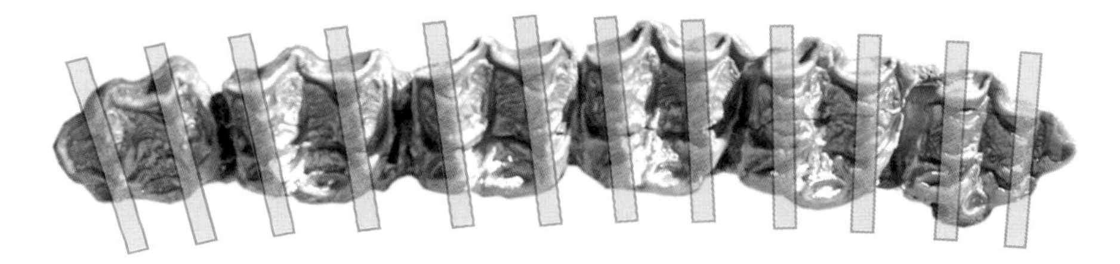

Figure 6.2. A view of the occlusal surface of right maxillary cheek teeth of *Hippotherium primigenium*. The transverse beams mark the orientation of the ridges of the secondary relief of a browser brought about by abrasion of the occlusal surface of a grazer.

lution is continual over time, and it is based on structures that are already present. Nothing develops out of nothing. Additionally, all the intermediate stages of evolution have to be able to survive in order for the organism to compete successfully. The German zoologist Günther Osche has put this idea into the following words: "In contrast with the constructions of man, organisms can never be closed for reconstruction." Also, their construction and their lifestyle have to correspond with the environmental conditions. For these reasons organismic innovations are limited in their evolution by internal and external conditions. For internal organs, such as the heart, it does not matter in what kind of natural environment its owner is living. The heart functions in the desert in the same manner as in the rain forest or in water. As a pump, it has simply to ensure a blood pressure sufficient for circulation. In such cases, the evolutionary development consists simply in optimizing the performance of the apparatus independently from the environment in which its owner is living.

Darwin recognized that to survive each species has to produce a sufficient supply of offspring. Selection and evolution are therefore inevitable, but what Darwin did not recognize is that they occur independently from environmental changes. The overproduction of offspring forces each species to explore and expand its biotope until its body construction limits its survival. In the end, each individual independently "decides" how far it may go. This way, the body construction influences the direction in which the species is going to develop. The improvement of the cost-performance ratio is also crucial for the direction evolution will take. Principally, the opportunity to change the function of an organ or organ system and/or to change the environment is available at any time, as long as the cost-performance ratio is satisfactory or improving. However, all intermediate stages have to function better or at lower costs than their predecessors. Otherwise, they, or more specifically their genes, are not positively selected.

What does cost-performance ratio mean in this context? Let us assume that all biological performances—from locomotion to propagation, from growth to heat production—need energy. And let us assume that for each organism there is only the amount of energy available that it takes up either directly from the sun or deep sea volcanoes ("smokers") or from the digestion of food. Under these conditions evolution—hence improvement of the cost-performance ratio of a construction—can only persist when there is an improvement in the energy balance of the organisms. This means that the relationship between energy uptake and energy expenditure has to improve. This relationship can improve either by functioning better at a constant cost, by reducing energy consumption for the same performance, or—ideally—by both. Under these preconditions, selectively neutral characters cannot exist because their development is tied to energy consumption and therefore part of the energy balance. These ideas about evolution are fundamental for its reconstruction. They were developed during the 1970s and 1980s by a group of young biologists and paleontologists at the Naturmuseum und Forschungsinstitut Senckenberg at Frankfurt am Main. They are known today as the Frankfurt theory of evolution (FTE).

What is the biological meaning of a character? To systematicists it is a feature characterizing a taxonomic unit, such as a species, a genus, or a family. Biologically, however, a character is a body structure or part of a body's construction that helps an organism to survive. Consequently, we have to have an idea as to what role such a structure fulfills within a functional system—and ultimately in the whole organism—before we can decide on the degree of its actual phylogenetic development and the direction of its evolution. More succinctly, a functional analysis has to precede any kind of phylogenetic interpretation and application of a character for systematics. What does all of this mean with respect to the phylogenetic development of horses?

In contrast to all other mammals, horses are characterized as one-toed ungulates that stand and run on the tips of their toes. How did they reach this position and why did they lose their side toes—two on the inner and two on the outer side—in the course of evolution? The generalized foot of a mammal consists of five toes, and this configuration is the same in bears and human beings. During walking, the foot touches down first with the heel (the calcaneum

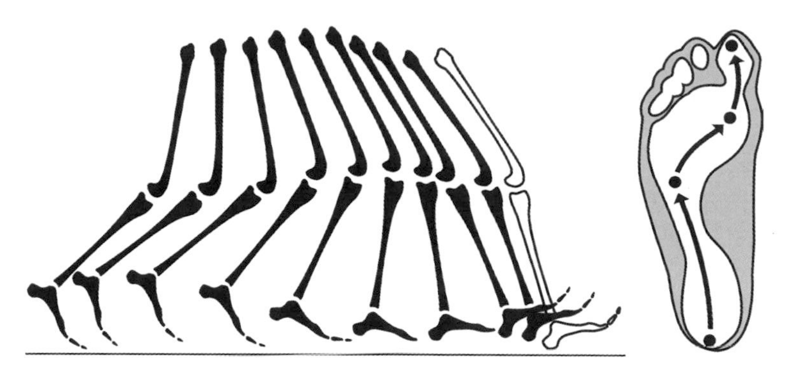

Figure 6.3. Humans are sole walkers. The human foot, like that of some other mammals, is pentadactyl. Once the human foot is at the tip of the hallux (big toe) it leaves the ground (*sequence at left*). The foot is then repositioned until the heel is placed on the ground. The arrows on the sole of the foot (*right*) show how the heel touches down first, rolls over the sole onto the ball of the foot, and then moves to the tip of the hallux.

or heel bone). After that it rolls off the outside of the sole until the ball makes contact with the ground. The toes, specifically in humans the big toe, or hallux, are the last part of the foot to make contact with the ground (Figure 6.3). Humans walk like bears, on the soles of their feet, although we walk on two legs, not four. The energy requirements needed to move on the tips of the toes are well known to ballet dancers, who use special shoes for this purpose. Horses, however, were not equipped with such support during their evolution. And dancing bears do not move on their tiptoes. They are notable simply because they dance. So why did horses evolve to not only stand but walk, and even gallop, on the tips of their toes? How did this process take place, and what were its advantages?

Let us begin with the advantages. Like everything in nature an organism's movements correspond to the rules of physics. In this context, the laws of pendulums, inertia, and levers play a role. A pendulum is an object that is able to rotate around an axis. Under the influence of gravity it moves periodically back and forth. The power of this movement increases with the mass and the length of the pendulum. On the other hand, a pendulum's swing means a permanent reversal of the direction of movement. This reversal becomes slower the larger the mass of the pendulum. In this context, the law of inertia—as recognized by the German astronomer Johannes Kepler (1571–1630) and later formulated by the English physicist and mathematician Isaac Newton (1643–1727)—plays a role. It states that an object stays at rest or in constant motion in one and the same direction, as long as it is not influenced by external power. In the case of a pendulum, gravity slows down the movement. Its effect is increased with the length of the pendulum. In the case of the limbs of horses, this means that their swinging becomes slower the larger their mass is and the closer it is situated to the end of the limbs.

As a result it is not surprising that the evolution of the limbs of horses was on the one hand aiming to increase the limb length in order to increase the length of each stride, and on the other hand to decrease inertia by decreasing the masses, particularly those closest to the end of the extremities. For that reason, the muscles in the lower parts of the limbs were replaced more and more by tendons. On the other hand, the large groups of muscles—or in other words the engines of locomotion—developed as close as possible to the trunk and to the axes of the shoulder and the pelvic joints. Their power was transmitted more and more throughout evolution by the tendons, which functioned as bundled up transmission belts and were guided by bandages of ligaments in order to have the main effect at the end of the limbs, pushing the animal forward. In this way, the legs' inertia became considerably reduced. In order to complete the system, the bones act in the form of poles as antagonists of the muscles and strings, transmitting the power to the ground (Figure 6.4). These physical and constructional circumstances and connections were identified by William Diller Matthew (1871–1930)—famous American investigator of the evolution of the horses—during the first half of the twentieth century.

With Matthew's explanation in mind, we can see that the change from sole to tiptoe walking led to an increase in the length of the pendulum of the limbs, and hence an increase in the stride length. Elongated extremities are clearly advantageous. And the lighter they become, the more reduced their inertia, and the better they are suited for locomotion. At a given speed the decrease in power necessary for its achievement results in an increase in perseverance for fleeing prey, such as horses. This is not so important for the chasing carnivores, which may take a rest if the hunting is not successful. For predators it is more important to be able to accelerate with a short burst of speed and to employ high limb maneuverability in order to react as soon as possible to the evasive actions of the prey. Moreover, for predators the limbs are used for locomotion and for holding onto the prey. Thus the locomotory apparatus of predators comprises heavy muscles and numerous bones near the ends of the limbs. This results in power and flexibility. The input of energy is correspondingly high. On the other hand, the strategy of fleeing animals is to exhaust the predators so they give up the chase. Predators have developed various strategies to contend with that situation. Cats (such as the cheetah) lie in wait for their prey in order to catch it after a short, high-speed sprint. Dog packs, on the other hand, chase prey to the point of exhaustion. For this reason dogs have evolved longer, more energy-saving extremities.

In horses, the poles (legs) should be as light as possible without losing stability. This is achieved by reducing the foot structure to only one supporting pillar, which is the middle ray of the foot. Here the laws of levers are playing a role. Levers are structures mostly in

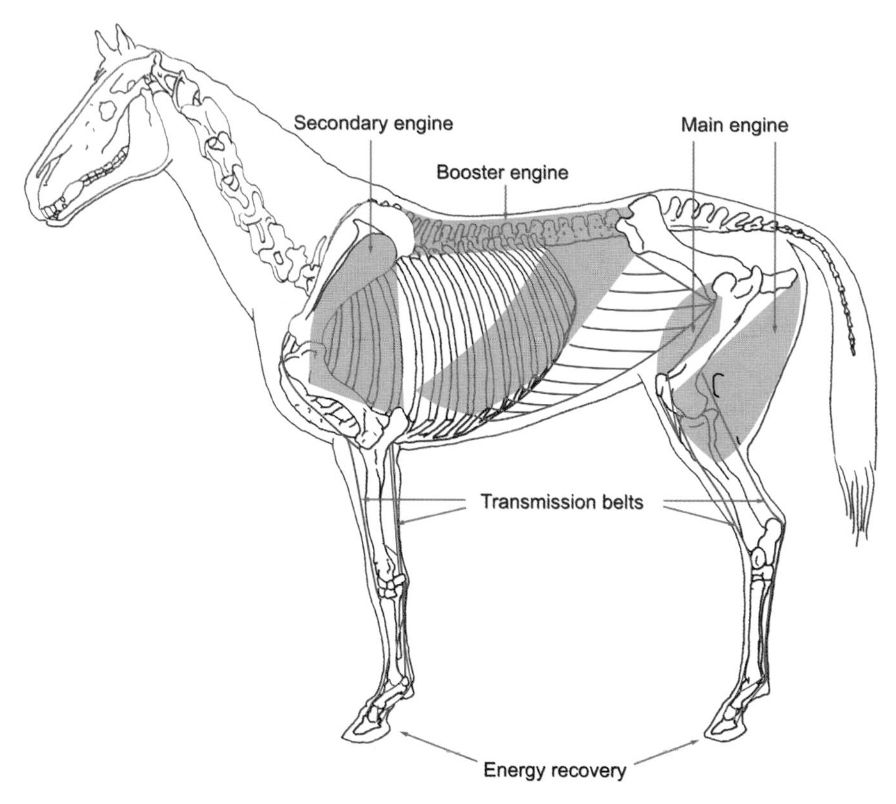

Figure 6.4. The evolution of horses brought an elongation of the limbs (to increase the stride length) and a decrease of inert masses, particularly near the end of the limbs. Heavy muscle groups in the lower parts of the limbs are replaced by much lighter tendons (*yellow*). Thick muscle groups situated near the trunk, shoulder axes, and pelvic joints (*pink*) developed into strong engines. Their power is transmitted to the tips of the toes by long tendons functioning rather like transmission belts. The tendons are guided by ligaments. In addition to the main engine in the back and the secondary engine in the front, horses occasionally use a booster engine consisting of dorsal and ventral musculature in the lumbar area. The bones (*white*) act as the poles of the system. (See also Figure 6.11.)

the form of a pole rotating around one constant axis. The law of the lever says: A lever is balanced when the torques of all the powers and loads are zero. For a simple straight lever the following equation is effective: Power divided by power lever is equal to load divided by load lever. Power lever and load lever correspond to the distances between their points of action and the axis of rotation. The longer the power lever the stronger the power, and the longer the load lever the heavier the load. These laws are crucial for understanding the evolutionary reduction of the side toes in horses. Side toes transmit their portion of the body load to the ground at longer load levers than the middle toe. If the load is focused on the middle toe, the lateral load levers are shortened and finally completely reduced. In that way, the whole locomotory apparatus is improved. It is no wonder that the evolution of the horses is characterized by a progressive reduction of the lateral toes, accompanied by a corresponding reinforcement of the middle toe (Figure 6.5). This development, however, is linked with restrictions on functions. Cats, for example, can use their five-digit paws not only for locomotion but also for defense, body care, and grasping and manipulating prey. But other than for locomotion, horses can only use their hooves for lashing out. As seen from the standpoint of a cat, the foot of a horse is a rather primitive construction. As seen from the standpoint of a horse, however, its functional limitations result in the highest form of specialization.

The swinging back and forth of the limbs is restricted by the plane of locomotion, which helps to save energy. Side or turning movements are largely avoided due to the increasing fusion of the ulna and radius in the forelimbs, and the shinbone (tibia) and calf bone (fibula) in the hind limbs (Figure 6.6). The tightened guiding rails of the articulations also help to restrict movements in the plane of locomotion. Thus the evolution of the horses' limbs is explained independently from the ecological conditions in the sense of construction optimization, and is instead similar to the Frankfurt theory of evolution (FTE). It is no wonder that this process of economization took

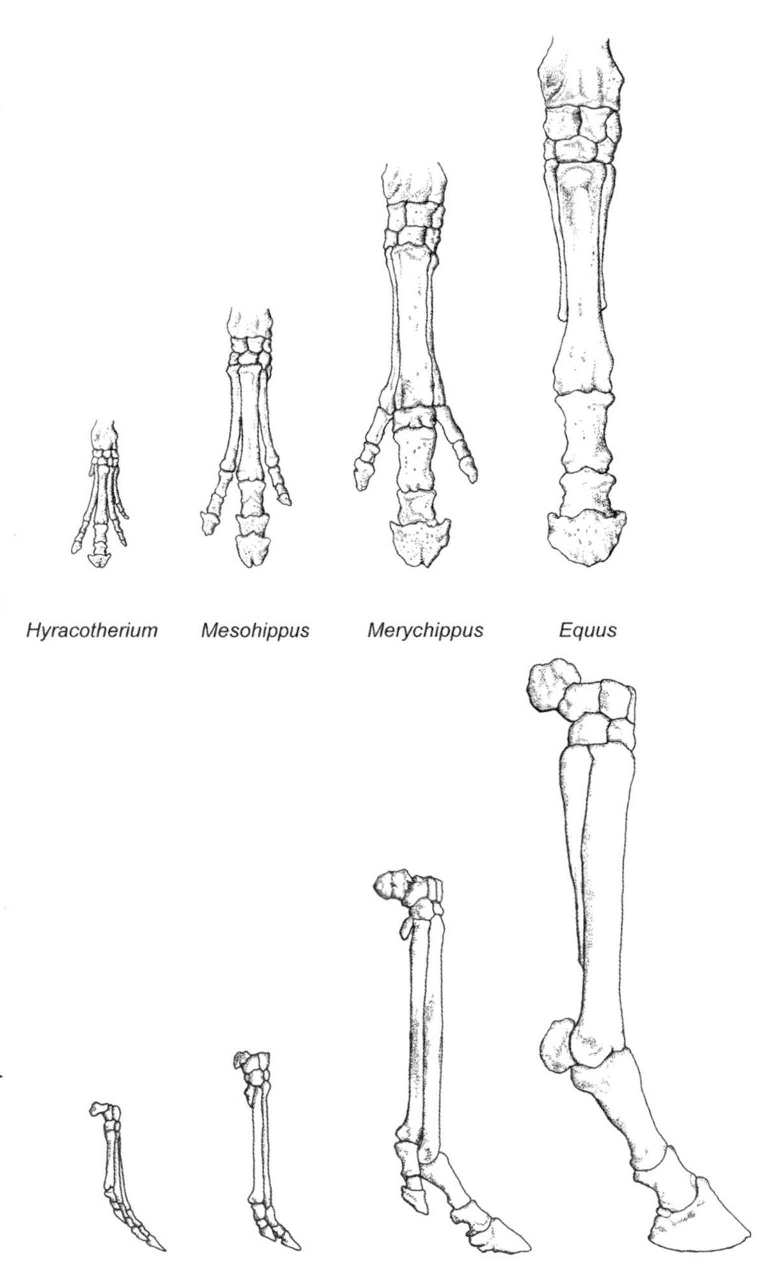

Figure 6.5. Evolution of the horse foot, which took place over 55 million years, is shown here in four stages: *Hyracotherium*, *Mesohippus*, *Merychippus*, and *Equus*. In the top row the forefoot is seen from the front; in the bottom row the hind foot is seen from the side. The evolution of horses is characterized by a progressive reduction of the side toes and a reinforcement of the middle ray. (See also Figure 6.10.)

place during most of the evolution of horses from the early Eocene up until the early Pliocene. It was not restricted to the transition from forest to steppe/savannah environments at the beginning of the Miocene or earlier.

Nonetheless, evolution cannot occur without an environment. Nor can organisms, including horses, live without an environment. Polydactyl (many-toed) feet were advantageous on soft, possibly even morassic, forest soils because they distributed body weight over a larger supporting area. In earlier times, it was not unreasonable for farmers to strap the hooves of their horses with large clogs in order to avoid their sinking into smooth meadows. One-hoofed horses are only practical on solid ground, such as savannahs and steppes. But such a change of environment cannot sufficiently explain the evolution of the horses during the whole range of time. At best, it is an additional aspect. The constructional developments made such a change of environment possible and not vice versa.

Up to this point, neither tiptoe walking nor its evolutionary development is explained. An evolutionary elongation of the extremities, a restriction of their movements on the plane of locomotion, and a reduction of the inert masses near their ends are all advantageous and necessary without tiptoe walking. Consequently, there must be another reason for its development. This special reason was discovered in 1942 by two North American zoologists, Charles Camp and Natasha Smith. Again it has to do with economization of locomotion. Camp and Smith recognized that certain marks on the un-

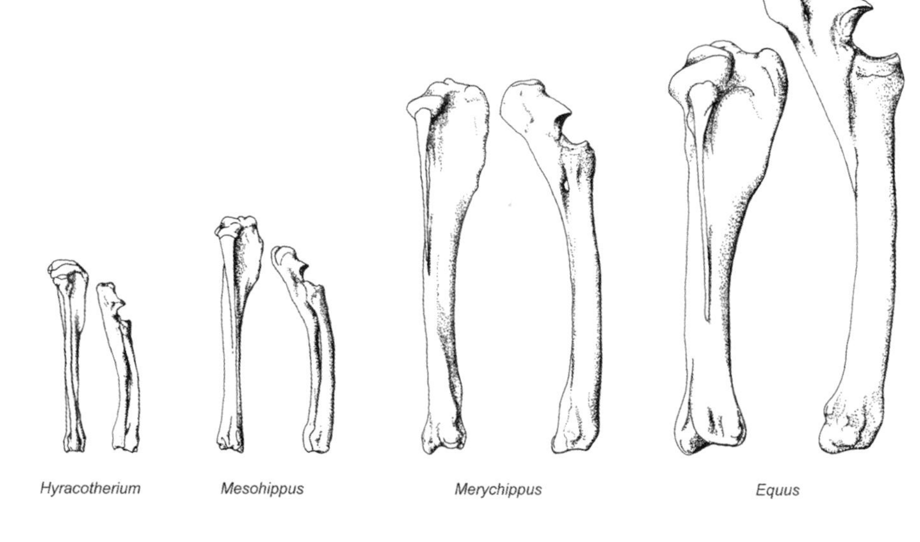

Figure 6.6. The back-and-forth motion of the limbs was restricted more and more in the plane of locomotion as horses evolved. Laterally directed movements would have considerably slowed down the horse's speed. The increasing fusion of the ulna and radius in the forelimb (*shown on the right in each pair*) and the tibia and fibula in the hind limb (*left in each pair*) led to the evolution from *Hyracotherium* through *Mesohippus* and *Merychippus* to *Equus*.

derside of the phalanges of the hands and feet indicate that the tendons and ligaments became considerably reinforced during the evolution of horses (Figures 6.7 and 6.9). What does this mean with respect to locomotion?

As already mentioned, the to and fro swinging of the limbs corresponds with the movement of a pendulum. Each footing, however, leads to a brake in locomotion until speed increases again by squeezing off the foot. As a consequence, a lot of energy is lost by being transferred into heat. Ligaments and tendons, however, are elastic. Therefore, they are able to store energy during footing in order to release it while squeezing off. This was what Camp and Smith discovered, namely that the construction of horses' feet corresponds to an energy-saving apparatus, or more appropriately an energy-retrieval apparatus, like some sort of afterburner. Now we know the reason for its specific construction. But how did this development take place in the course of evolution, and what were the better and better functioning transitional stages between sole and tiptoe walking?

In order to find an answer to these questions, it is worthwhile to look at other ungulates of similar build, for example, the tapirs. Tapirs, like horses, belong to the perissodactyls. With regard to the development of their foot construction, however, they stayed halfway

Figure 6.7. Early on in the evolution of horse locomotion, the horses raised up on the tips of their toes with the help of pads—which originally served as shock absorbers—on the underside of their feet. Later on, these inert masses were reduced as tendons and ligaments on the underside of the phalanges became strong enough to stabilize the foot. (See also Figure 6.9.) Additionally, the elastic tendons and ligaments save kinetic energy when they touch the ground (the energy is released again at lift off).

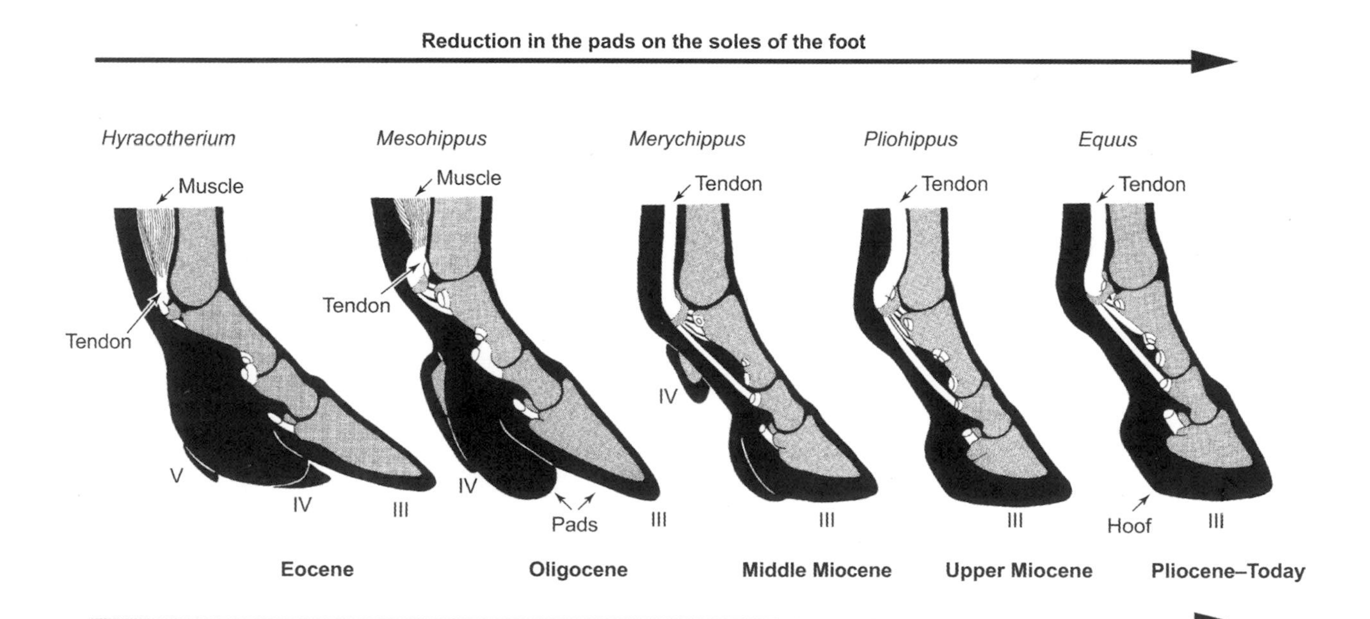

Figure 6.8. Tapirs serve as "living fossils." Note the underside of the hind foot of this mountain tapir (*Tapirus pinchaque*), which shows a clearly visible pad in the shape of a heart. When compared to a horse's foot, a tapir's foot might be considered only halfway developed. The middle ray of a tapir's foot is reinforced, but an inner and an outer toe are still present. On the front limb, tapirs even have a fifth toe, similar to the dawn horses of the Eocene. The pads on the soles of the tapir feet permit them to walk on tiptoe.

behind modern horses. It's true that they also have reinforced the central metapodia. But just as with the dawn horses of the Eocene, they still have an inner and an outer toe on the hind feet, while in front they display one additional toe (an external one, the fifth toe). On the other hand, tapirs no longer walk on their soles like humans and bears; they walk on the tips of their toes so that the metapodia and proximal phalanges don't touch the ground any more. How did they arrive at that position? Tendons and ligaments on the underside of their feet are still not strong enough to bear the body's weight. There are, however, pads on the underside of the feet that make elevation on tiptoes structurally possible (Figure 6.8). It is assumed that such pads served as shock absorbers in the dawn horses of Eocene times, as is the case with many mammals today. These pads increased in size during evolution in order to improve shock absorption. At the same time, the limbs became elongated as the size of the pads increased, thus improving locomotion. Using this as a model, horses achieved tiptoe walking continuously, accompanied by a steady economization.

Normally, soft body constructions such as foot pads are not fossilized. Nevertheless, we have to postulate their existence in dawn horses at least until the beginning of the Miocene epoch, around 22 million years ago, because they are the only construction that makes such a continuous transition from sole to tiptoe walking understandable. When tiptoe walking is achieved, the pads at the end of the extremities are reduced as inert masses. This reduction, however, is only feasible if the ligaments and tendons on the underside of the feet become strong enough to support sufficient body weight. At the same time, they improve the energy balance of the animal due to their elasticity. Although this evolutionary development is not fossilized, the fossil record nonetheless demonstrates this. The ligaments and tendons, which are not preserved directly, can actually be reconstructed from the marks they left behind on the fossilized underside of the foot bones in the course of their evolution (Figure 6.9).

In light of this background—that it took from the early Eocene until the beginning of the Pliocene epoch ca. 5.5 million years ago for horses to reduce their side toes so that they no longer touched the ground—the evolution of horse locomotion becomes understandable. Much of this was apparent at the beginning of the Miocene, about 23 million years ago (Figure 6.5). For a long time, the evolution of horse hooves was considered proof that useless, nonfunctional characters were dragged along by evolution. It was the German mammalian paleontologist Heinz Tobien (1911–1993) who analyzed

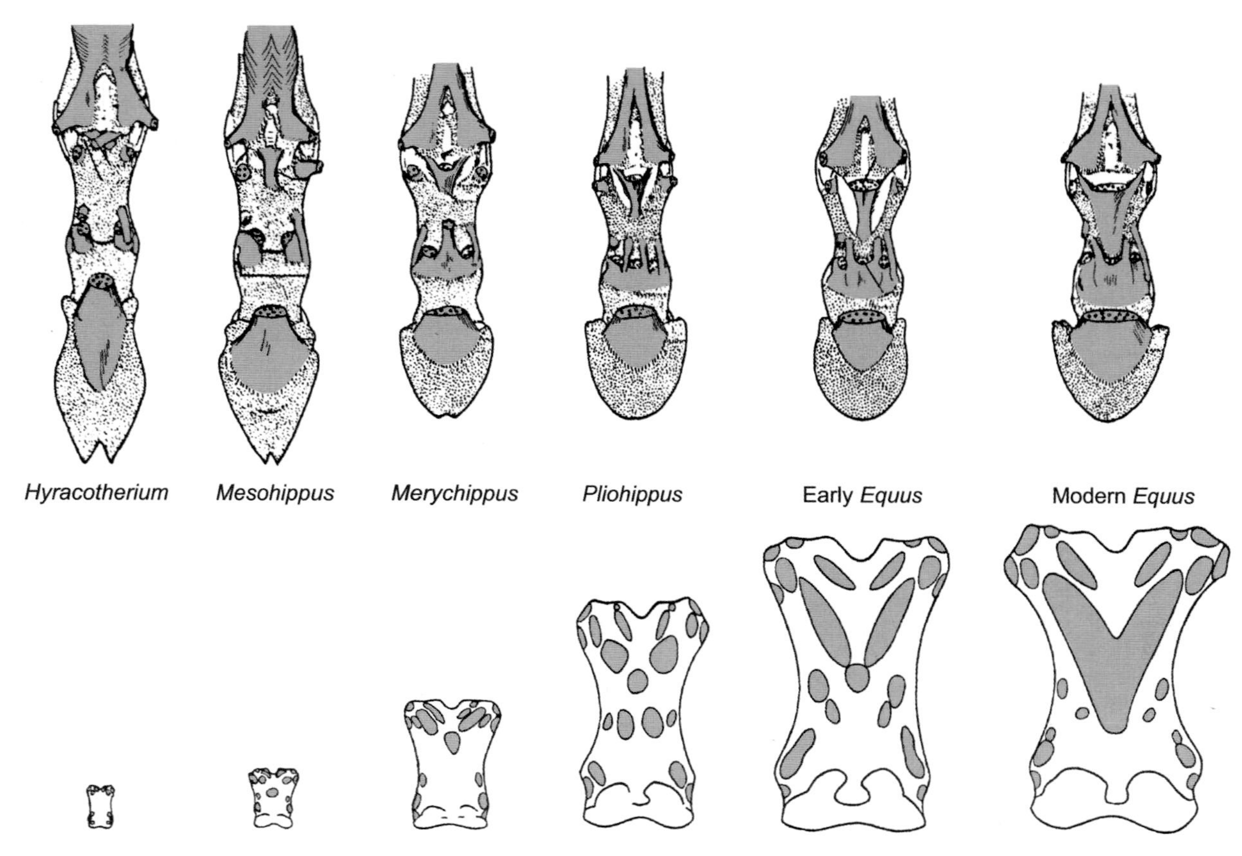

Figure 6.9. Soft tissue is not usually preserved in fossils. But in the case of Paleogene dawn horse fossils (*bottom row*), tendons and ligaments (*yellow*) left behind marks on the underside of the phalanges that increase in size throughout evolution as shown in these illustrations. The presence of sole pads, ligaments, and tendons is assumed for these horses because their feet simply would not function without them. Sole pads paved the way for a transition from sole to tiptoe walking, which also improved the animal's energy balance.

and reconstructed the foot skeleton of 10–million-year-old hipparionine horses from the Höwenegg. He demonstrated that the side toes were in no way useless, but on the contrary absolutely necessary for survival. During phases of extreme strain, they helped support the apparatus of ligaments and tendons before the feet had evolved to be strong enough to absorb such stress (Figure 6.10). Only when the ligaments and tendons became strong enough to bear the full body weight, even during phases of extreme stress, could the side toes be reduced completely. The only exceptions, which are still present today, are the relics of the splint bones. It was in this way that the horses evolved into modern-day one-hoofed ungulates. This process was completed during the Pliocene about 4 million years ago during the development of the genus *Equus*. With *Equus*, horses arrived at the climax of their evolution: They stood and ran on the tip of one toe without any support from pads or side toes.

In order to understand the development of the whole locomotory apparatus, it is necessary to return once more to the engines behind the movements. There are three. The main engine is the thick package of the gluteus musculature situated between pelvis and femur (Figure 6.4). This engine transmits power to the hind limbs. Adding to that is a similar, although weaker, package of muscles around the shoulder blade (scapula). This transfers power to the anterior limbs. The third engine does not exist in the limbs. It is located on the lumbar region of the vertebral column. The dorsal or back musculature acts together with the lumbar vertebral column and the ventral musculature to facilitate bow flexing and stretching at short intervals. This engine comes into play as some kind of supplementary aggregate, so to speak, during fast locomotion, such as galloping. Its role is quite visible during horse races, when the jockeys stand in the stirrups (Figure 6.11). Jockeys do this not only for better streamlining but also to provide more room for the recurrent ups and downs of the horses' vertebral column.

Naturally, the development of the third engine played a role during the evolution of horses. The dawn horses of the early Eocene still had a vertebral column that was highly arched (Figure 6.12a), similar to modern duikers and muntjacs (Figure 6.12b). This column was stretchable in the lumbar area. On the other hand, the complete skeletons of *Hippotherium primigenium*—unearthed from 10-million-year-old lake deposits at the Höwenegg near Lake Constance—showed a vertebral column that was already considerably stretched out (Figure 6.13a), although not to the extent it is in modern-day horses (Figure 6.13b). The third engine probably became more and more important as the horses ventured into the expanding and differentiating biotopes of steppe and savannahs at the beginning of

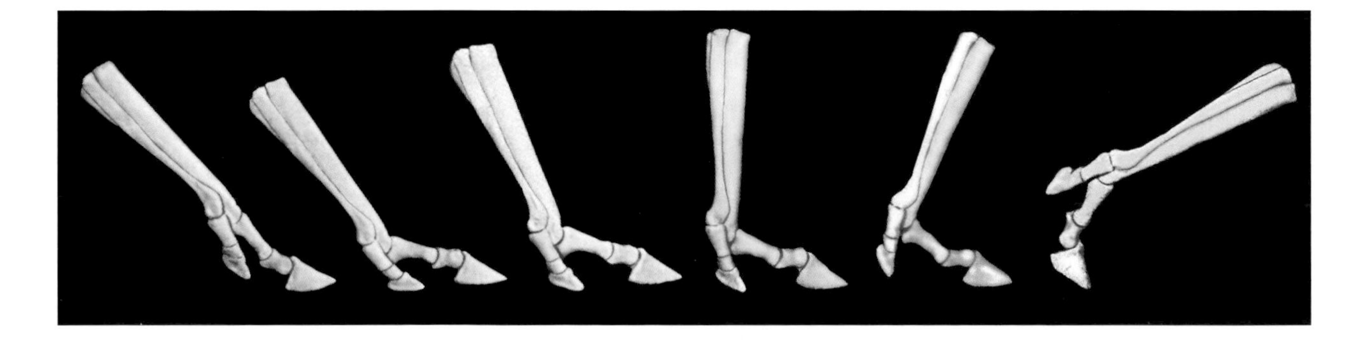

Figure 6.10. Function of the side toes during movement in *Hippotherium primigenium* (which lived approximately 10 million years ago) from touch down on the left to lift off on the right. During phases of extreme strain, the side toes acted as additional supporters, preventing the crashing of the foot and tearing of tendons and ligaments.

Figure 6.11. Horse race at Ellerslie Park, New Zealand. The back musculature of a horse's lumbar region forms a third engine that operates in addition to the shoulder and pelvic musculature (see Figure 6.4). This auxiliary engine only comes into play during times of very fast locomotion, such as a horse race. When jockeys stand up in the stirrups, they are allowing as much space as possible for the flexing and stretching of the horses' auxiliary engine.

the middle Miocene. Here, fast locomotion, such as galloping, which was almost impossible in the undergrowth of Eocene rain forests, became more and more important.

Overall, in the course of evolution, the dawn horses of the Eocene developed step by step from small and agile bush slippers into highly specialized runners. Does this mean that they became faster? This question is difficult to answer because there were no eyewitnesses to measure the speed in previous times. As is frequently the case in paleontology, we have to rely on clues. The great mammalian paleontologist George Gaylord Simpson (1902–1984) once compared the proportions of the early Eocene North American *Hyracotherium* (*Eohippus*) with those of a greyhound (Figure 6.14). In his opinion, the proportions were quite similar. Simpson concluded that dawn horses were about as fast as modern-day dogs. Coyotes have been measured at top speeds of 64 kilometers per hour. If we transfer that to the dawn horses of the Eocene, it means they were almost as fast as modern-day horses that reach top speeds of 70 to 72 kilometers per hour.

What does this mean with respect to evolution? Were all the specializations for running more or less in vain? Was it only chance that governed these changes? What was the effect of selection? Was there any?

If we assume that Simpson's analogy is right, there is an explanation for the high specialization of horses as runners that is convincing and also amazing in the light of evolutionary theory. It re-

Figure 6.12. (a) Early dawn horses of the Eocene had a highly arched backbone that was only stretchable in the lumbar region; (b) modern-day muntjacs have a similar build.

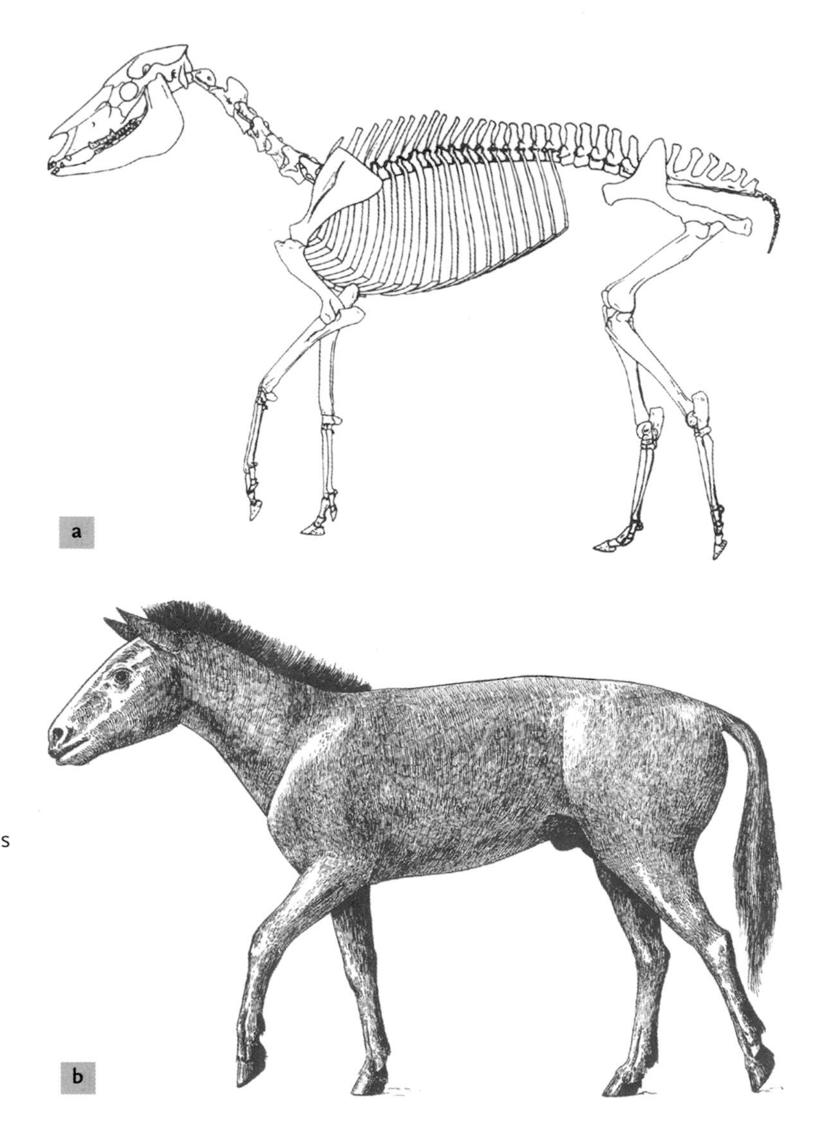

Figure 6.13. (a) Complete skeletons of *Hippotherium primigenium* from the Höwenegg (near Lake Constance) display a slightly curved dorsal spine similar to that of modern red deer; (b) an artist's rendering of what *Hippotherium primigenium* looked like.

sults from the tendency of horses to become bigger during evolution. This is also known as Cope's rule. Despite certain exceptions, there is generally a phylogenetic size increase. But why? Of course, it makes sense that an impressive body size would be an advantage both in defense against enemies as well as during mating competitions among males. There is, however, another much more fundamental advantage. As physiologists determined, the metabolism of larger organisms is more economical than smaller ones. This corresponds to the business world. The larger a factory or company becomes, the more economically it can work. In organisms, however, there is a special condition. While body size increases by a cube function (x^3), the strength of musculature corresponds with its cross section, which means that it increases only by a square function (x^2). Therefore, larger animals either become slower or they specialize or economize their locomotory apparatus in order to keep up with the speed of the pursuing carnivores or the pursued prey respectively. Specifically, fleeing ungulates like horses not only need to reach the speed necessary to escape but also have to maintain this speed for some time without losing too much energy.

Elephants and their ancestors, for example, chose the first option. They became quite slow but at the same time too big for predators (except for humans) to attack them successfully. For horses, the second alternative was adopted. Their evolution aimed at the economization of speed. Horses compensated for the negative consequences of becoming larger by maximizing the specialization of their locomotory apparatus. This is the reason why horses became such highly specialized runners. Of course such a development was not adopted deliberately, but was an outcome of natural selection. Animals that became slower with increasing body size fell prey more easily than those that were able to reach and maintain the speed necessary to escape.

Does the fossil record offer any information about the locomotion of horses throughout time? Are there any tracks that could be related to certain fossil species? And if so, what do they tell about the modes of locomotion and speed? Such tracks do exist, although they are very rare. Unfortunately, they do not tell us very much with respect to top speeds. One example from the famous volcanic ashes of Laetoli (East Africa) is about 3.5 million years old (Figure 6.15). In these layers the earliest tracks of the bipedal australopithecines were discovered. Next to them are tracks of various animals, among which are tracks of an adult and a juvenile horse, probably a mare and a foal

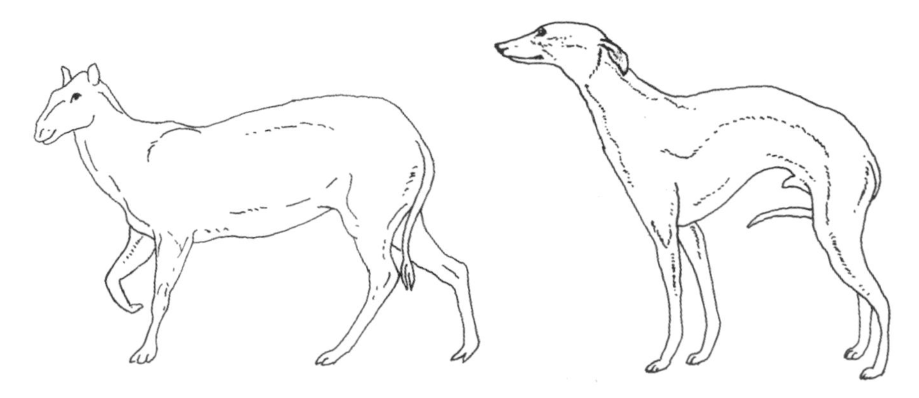

Figure 6.14. The famous mammalian paleontologist George Gaylord Simpson compared the proportions of the North American *Hyracotherium* (*Eohippus*) with those of a greyhound. He concluded that the dawn horses had a top speed of about 64 kilometers per hour and were almost as fast as modern horses.

of the three-toed *Hipparion*. Analyses showed that these early horses performed a so-called running walk on the slippery ground. With this mode of locomotion, one foot is always planted on the ground for stability. The speed was calculated at 6.5 to 15 kilometers per hour, which means that it was far from the speed achieved during galloping. Interestingly, the side toes still touched the ground. This confirms the assumption that the side toes functioned in this evolutionary stage for additional support during phases of strain, in this case on a tough substrate.

To what extent energy-saving played a role during the evolution of the horses becomes clear in context with the development of the locomotory apparatus. Surprisingly, energy can also be saved while standing. As the North American researchers Hermanson, MacFadden, and Hildebrand have shown, a certain arrangement of muscles on the shoulder and the knee joint provide the potential for these articulations to be clicked into place in order to minimize energy consumption by muscle activities (Figure 6.16). To what extent biomechanical factors fundamentally determine phylogenetic development and evolution of organisms becomes visible in the context of the economization of the locomotory apparatus, as well as with that of the chewing apparatus and the construction of the skull as a whole. It is not then the environment that determines the evolution of an organism and its organs. An organism is an extremely complicated structural system of a variety of components working together in many different ways. So it is no wonder that processes of economization take place within the body construction itself with respect to the development and cooperation of various organs. Examples of this are the development of the chewing apparatus and the skull, as well as the interplay between chewing apparatus and digestive system.

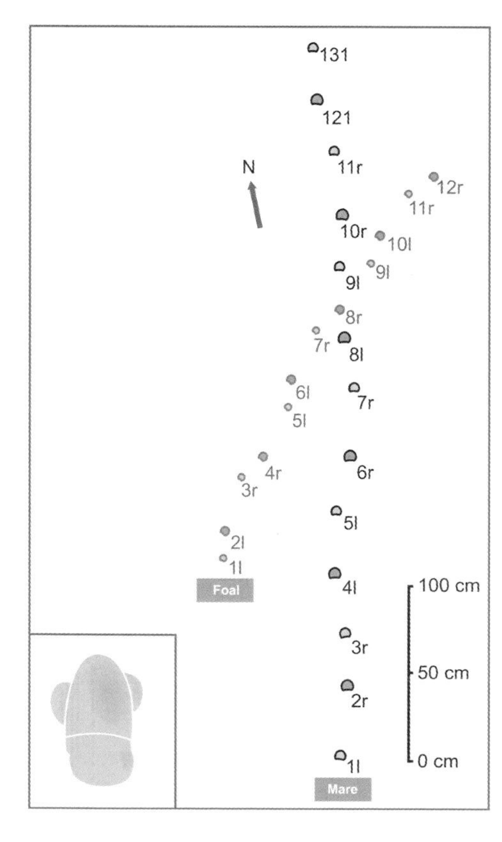

Figure 6.15. Illustration of fossil tracks of horses from about 3.5 million years ago. The tracks were excavated from volcanic ash near Laetoli in East Africa. They were made by a foal and an adult, presumably a mare, of the three-toed *Hipparion*. Odd numbers identify front hooves and even numbers, hind hooves; r = right and l = left. Analyses have shown that these early horses performed a so-called running walk in which one foot remained on the ground at all times. Their speed was calculated at 6.5 to 15 kilometers per hour. Isolated imprints (*inset*) indicate that the side toes touched the ground in order to avoid overstretching of the toes.

On the skull of a modern-day horse, the long stretched out facial part in front of the eyes is most conspicuous (see Figure 2.2). There are two causes for that elongation. One is the development of the nose, which is important for a herbivore that is unable to control its food intake visually. The other is immediately behind the mouth and in front of the digestive tract, the location of the food-processing chewing apparatus. While the body increases in cubes (x^3), the masticatory surface only increases in squares (x^2). Therefore it becomes necessary in the course of evolution to disproportionately enlarge the chewing apparatus. Initially this was achieved by molarization of the premolars (Figure 6.17). Further increases of the masticatory surface became necessary with the transition to grass food. It is essentially phytoliths, hard crystals of calcium oxalate which the plants developed in defense of grazing ungulates, that together with sand taken up from the ground led to an accelerated rate of tooth abrasion. In order to solve this problem, one masticatory surface was piled on top of another. This way the masticatory surface increases enormously over time and the height of the tooth crowns (hypsodonty) also increases at high rates (Figure 6.1). This in turn leads to spatial problems in the facial part of the skull. The chewing apparatus takes up so much space that the eyes, which were originally situated much farther in front, are pushed back behind the tooth rows. On the other hand, the part of the dentition that is plucking off the grass—the series of incisors—is stretched forward in the direction of the food resources, while the grinding and chewing teeth are shifted backward, close to the articulation of the jaws where the pendulum conditions are better. As a result, a gap or diastema develops between the incisors and the premolars. Humans use this gap for the snaffle in domesticated horses (see Figure 2.2).

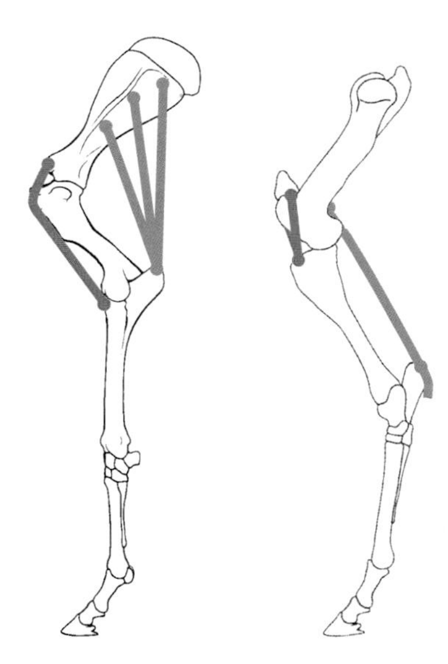

Figure 6.16. The importance of energy saving is demonstrated by the fact that horses even save energy while standing. An arrangement of antagonistic muscles balances the forces in the shoulder (*left*) as well as the knee joint (*right*). These joints can be clicked in place to minimize the energy consumed by muscle exertion.

A further change in the chewing effect is achieved by increased folding—an elongation of the enamel edges—significant for processing grass and other food (Figure 6.1). Crown height (hypsodonty) and enamel folding increase almost exponentially when the horses make grass their main food source. Nevertheless it is certainly wrong to consider these two developments as the only constructional preconditions for food change. How the food is acquired is important, as is how it is chopped up and ground. Therefore the preconditions of digestion are important (see Chapters 2 and 9). Grass contains a lot of cellulose. Without the ability to digest this nutritional substance, any transition to grass food would be useless. Consequently, the ability to digest cellulose would have been the first prerequisite before the ancestors of our horses could become grazers. A problem with respect to a grass diet exists because horses, as well as other mammals, do not possess the endogenic enzymes that enable them to digest cellulose.

As usual, natural selection found a way to solve the problem. In this case, bacteria are used to produce the enzymes necessary to digest cellulose. For this, bacteria become housed in the cecum, which is especially enlarged as a fermentation chamber. The enlarged cecum of the dawn horses from Messel indicate (see Figures 5.7 and 5.13) that these animals were already cecum fermenters 47 million years ago. This is true despite the fact that investigations of gut contents have shown that their nutrition did not yet consist of grass, but rather leaves, specifically laurels (see Figures 5.14 and 5.15). In a few cases seeds have also been found, particularly those of grapes (see Figure 5.16), suggesting that these dawn horses were also occasionally feeding on fruits. In these cases, however, it is not the food itself that was preserved, but that which was not digestible.

Techniques exist that allow researchers to check the kind of nutrition consumed even without the quality of gut content preservation found at Messel. Bushes and trees take in carbon by way of C3 photosynthesis, while most of the grasses prefer C4 photosynthesis.

Figure 6.17. The principal evolutionary development of equid dentitions about 33 to 53 million years ago is demonstrated by a comparison of the maxillary cheek teeth of *Hyracotherium* (*bottom*) and *Mesohippus* (*top*). In addition to an increase of size (not shown), the development consisted mainly of a molarization of the premolars (anterior cheek teeth) and the morphological transformation of bunodont (bumpy occlusal pattern) into lophodont tooth pattern (occlusal ridges).

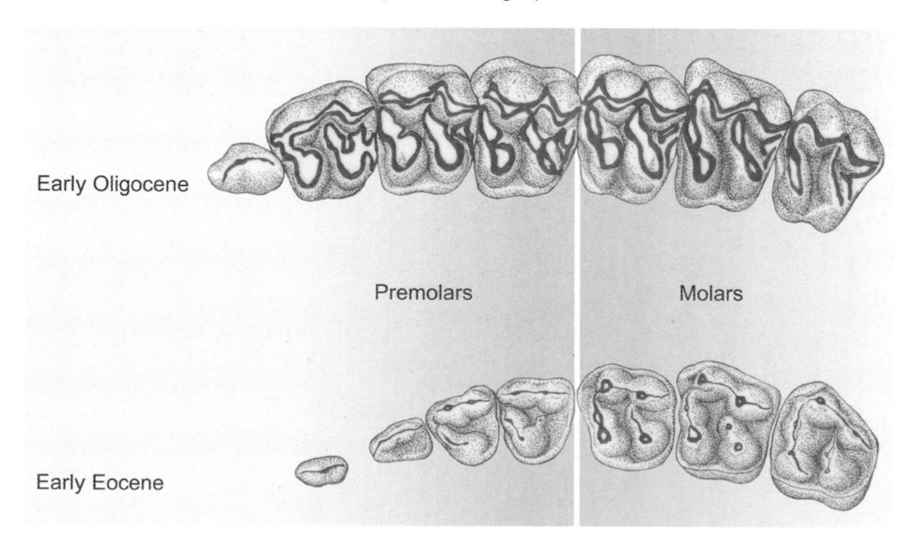

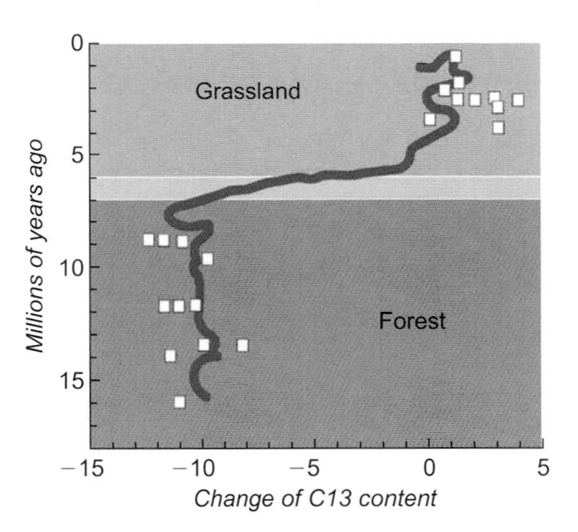

Figure 6.18. The change from forest to grassland about 6 to 7 million years ago in Pakistan was mirrored by a change in the content of carbon isotope ^{13}C in fossil soils (*gray curve*) and the ^{13}C found in the enamel of teeth (*open squares*) from the corresponding age. The main shift occurs at exactly the same time interval (*gray barrier*) as the changes in the shales of oceanic microfossils.

The syntheses differ from each other by their ratios of carbon isotopes $^{13}C/^{14}C$. C3 plants show ^{13}C portions of about 23 to 24 percent; C4 plants have only 9 to 17 percent. With the uptake of the appropriate amount of plant nutrition, the ^{13}C portion in the skeleton of these animals increases about 12 to 15 percent. Analyses of skeletal material distinguish browsers from grazers on the basis of their ^{13}C values. This holds true for living mammals as well as for fossils. Today such analyses are mostly applied to tooth enamel because this material is much denser than bone and therefore less prone to becoming contaminated or degraded. Since the late 1980s mammalian paleontologists such as Bruce McFadden at Florida State University have been able to use this as a tool to check the classic hypothesis of increasing crown height (hypsodonty) in horse teeth with the expansion of open grassland. Such an analysis was made with sequences of fossil horse teeth coming from Miocene sections of the Siwaliks of Pakistan. The result was extremely surprising. There is indeed an observable increase of ^{13}C in the tooth enamel, but this occurs not at the time of sudden increase of hypsodonty about 20 million years ago but much later, only 6 to 7 million years ago (Figure 6.18). A ^{13}C increase in fossil soils and in the shale of marine microfossils indicates a fast expansion of C4 grasses at the same time. How then is the increase of crown height explained, when it appears in Eurasia 4 to 5 million years earlier when the first hipparionine horses emigrate from North America?

The direct relationship between increasing crown height and grass food also became ambiguous when another aspect was investigated. The Swedish paleobotanist Caroline Strömberg determined that phytoliths—that is, grassland—appeared in the American West at least 10 million years earlier, at the beginning of the Oligocene, a long time before high-crowned cheek teeth developed in horses. Phytoliths found in the coprolites of some dinosaurs indicate that these animals were feeding on grass during the Cretaceous. All of this means that there is no temporal and also no causal connection between the origin and expansion of grasslands and the development of hypsodonty during the evolution of the horses. Increasing body size also cannot be the reason for hypsodonty because there were large horses in the late Miocene of North America and China—such as *Hypohippus, Megahippus,* and *Sinohippus*—with enlarged but not hypsodont cheek teeth (Figure 6.1). On the other hand, there were equids with extreme hypsodont dentitions—such as the North American *Nannippus* or the North African and Spanish *Hipparion periafricanum*—that were markedly small.

Obviously, the relationships are more complex than we have imagined. New research results from the American West—Colorado, Nebraska, Wyoming, Montana, and Idaho—indicate that the grasses diversified there at the beginning of the Oligocene, about 34 million years ago. They became dominant, however, no earlier than 7 to 11 million years later, at the transition from the Oligocene to the Miocene. In other regions, such as the Siwaliks, the change from forest to grassland occurred still later. Basically, in the manner of the Frankfurt theory of evolution these results call into question whether there is a close connection between environmental change and evolution at all, as Darwin assumed in his evolutionary theory. It seems it is not the environment that forces the development of adaptations by way of natural selection. On the contrary, it is the change of body construction in order to improve the energy balance that permits some organisms to venture into environments not inhabited by them before. In the same manner organs also take over new functional roles.

During the middle Miocene, about 13 to 18 million years ago, horses not only developed high-crowned cheek teeth but also diversified in a remarkable way. The classification of genera and species, mirrors, more or less, the basic attitude of the systematicists. There are splitters, who tend to make particularly fine, sometimes excessive differentiations, and there are lumpers, who tend to make broad, sometimes even crude, classifications. In the middle Miocene, the number of horse genera increases from 5 to no fewer than 13. After that, it decreases again, first gradually, then abruptly up until the Plio-Pleistocene, when only 1 genus remains, the modern-day *Equus* (Figure 6.19). Does the reduction of genera correspond to the reduction of the number of different biotopes? It has been only 6 to 7 million years since the horse's ancestors were restricted to modern steppes and savannahs dominated by C4 grasses that were finally inhabited by only one genus of horses. Or was the increasing selection at the beginning of the ice age due to a multitude of more or less abrupt climatic oscillations (see Chapter 10)? During this same time, other mammals also show a considerable decrease in biodiversity. On the other hand, the glacial climatic oscillations begin much later than the decrease in biodiversity. Therefore, the decrease in biodiversity seems to be better explained by a shrinking of a multitude of biotopes due to a general change of climate and vegetation.

And what do the genes tell us? This question is particularly pertinent in light of the success of molecular genetics in criminology. The following can be said: The oldest DNA known for fossil horses is "only" 53,100 years old. It was discovered in a member of the genus *Equus* from Siberia. Together with the data about more recent Pleistocene and modern horses, it has led to a phylogenetic tree of Pleistocene equids that is in some ways revolutionary (see Chapter 10). However in spite of such successes, one has to be careful not to expect miracles from this method. Molecular genetics cannot reach very far back in the earth's history because DNA does not necessarily stay viable over the long haul. In addition, molecular genetics, like morphology, offers relational information based on similarities, and these are a function of probabilities. Molecular genetics is an extension of morphology into the molecular realm, so to speak. Up to now, not one fossil animal has been reconstructed using molecular genetics. Animals without living relatives would remain forever unknown if we relied only on molecular genetics. Nothing would be known of dinosaurs. Nothing would be known of the dawn horses of the morning cloud if paleontology—the science of life in the geologic past—had not unearthed the relevant fossil documents and then interpreted them. Molecular genetics is at its best with respect to phylogenetic relationships of animals that are still living today, such as asses, half-asses, zebras, and horses. Therefore, it helps to bring to light the phylogenetic development of horses during the Pleistocene. The much larger part of the horses' phylogenetic tree, however, which extends far back in the earth's history, remains beyond the reach of molecular genetics. The evolutionary development of horses that took place before the ice age can only be reconstructed using fossil discoveries analyzed with the methods of constructional and functional morphology.

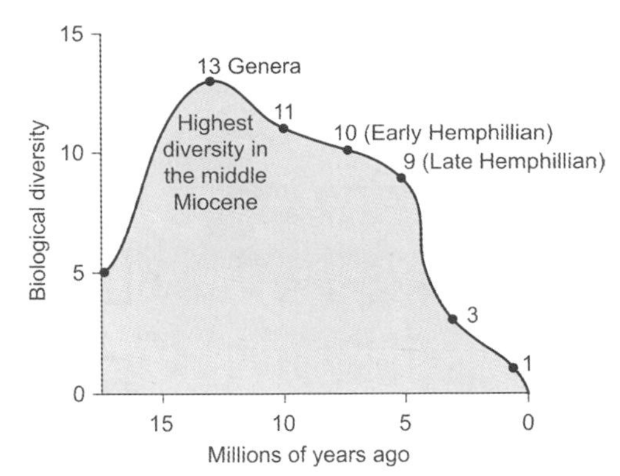

Figure 6.19. During the Middle Miocene, about 13 to 18 million years ago, the North American horses showed an extraordinary diversification. In the course of this process the number of genera increased from 5 to no fewer than 13. Diversity then decreased slowly until the beginning of the Pliocene, about 5 million years ago, leaving *Equus* as the only genus. The explanation for this increase and decrease could be that with the increasing expansion of grasses during the Middle Miocene a multitude of different biotopes developed that were later reduced to only savannahs and steppes due to climatic and environmental changes at the beginning of the Pliocene.

as the wild-living representatives are considered). In 1876, with respect to the increasing distribution of grasses during the Tertiary, he wrote that "consequently it was most likely food, which caused the depicted modifications in tooth construction that led to the large, pillared, permanently growing teeth of modern ungulates from comparatively fine-rooted teeth of the Eocene and Miocene forms."

Kowalevsky envisioned horse progression the same way T. H. Huxley did (although he did not specifically refer to him): *Palaeotherium* < *Anchitherium* < *Hipparion* < *Equus*. It was a purely Old World horse progression. Only with *Hipparion* did he add in brackets, "perhaps *Meryhippus* Ld." Concerning the knowledge of the North American forms, he bemoaned the fact that Joseph Leidy (1823–1891) of Philadelphia had not illustrated a single postcranial skeletal element of fossil horses in his publication on the "ancient fauna of Nebraska."

In the absence of a sufficient exchange of information, two completely different horse progressions developed on either side of the Atlantic. The first scientific expedition into the western territories of the United States, most notably carried out by Leidy, produced rich finds of fossil horses. This material was processed primarily at Yale University by Othniel Charles Marsh (1831–1899) and in Philadelphia by his bitter rival Edward Drinker Cope (1840–1897). Leidy himself never tried to reconstruct a horse progression with his discoveries, although he definitely recognized that his finds were relatives of horses.

For their part, Cope and Marsh carried out widespread expeditions in the Wild West. As a result of the expeditions, in 1874, Marsh published a North American evolutionary tree of horses that proceeded from *Orohippus* (Eocene), to *Mesohippus* (Oligocene), to *Miohippus* (*Anchitherium*; Miocene), to *Protohippus* (*Hipparion*) and *Pliohippus* (both from the Pliocene), and finally to the Pleistocene and modern-day *Equus*. This horse progression displayed a gradual reduction of the lateral toes, of the forearm (ulna) and calf (fibula) bones, as well as an increase in the crown height and complexity of the occlusal surface of the molars (Figure 7.3).

In 1876, Huxley traveled on a lecture tour to the United States and visited Marsh in New Haven. While he was there, he admitted that the North American fossil horses corresponded with the postulated evolutionary history much better and documented this development more completely than the European genera. Huxley consequently modified his lecture. The progression of horses with its significant features was established at that time much as it is today.

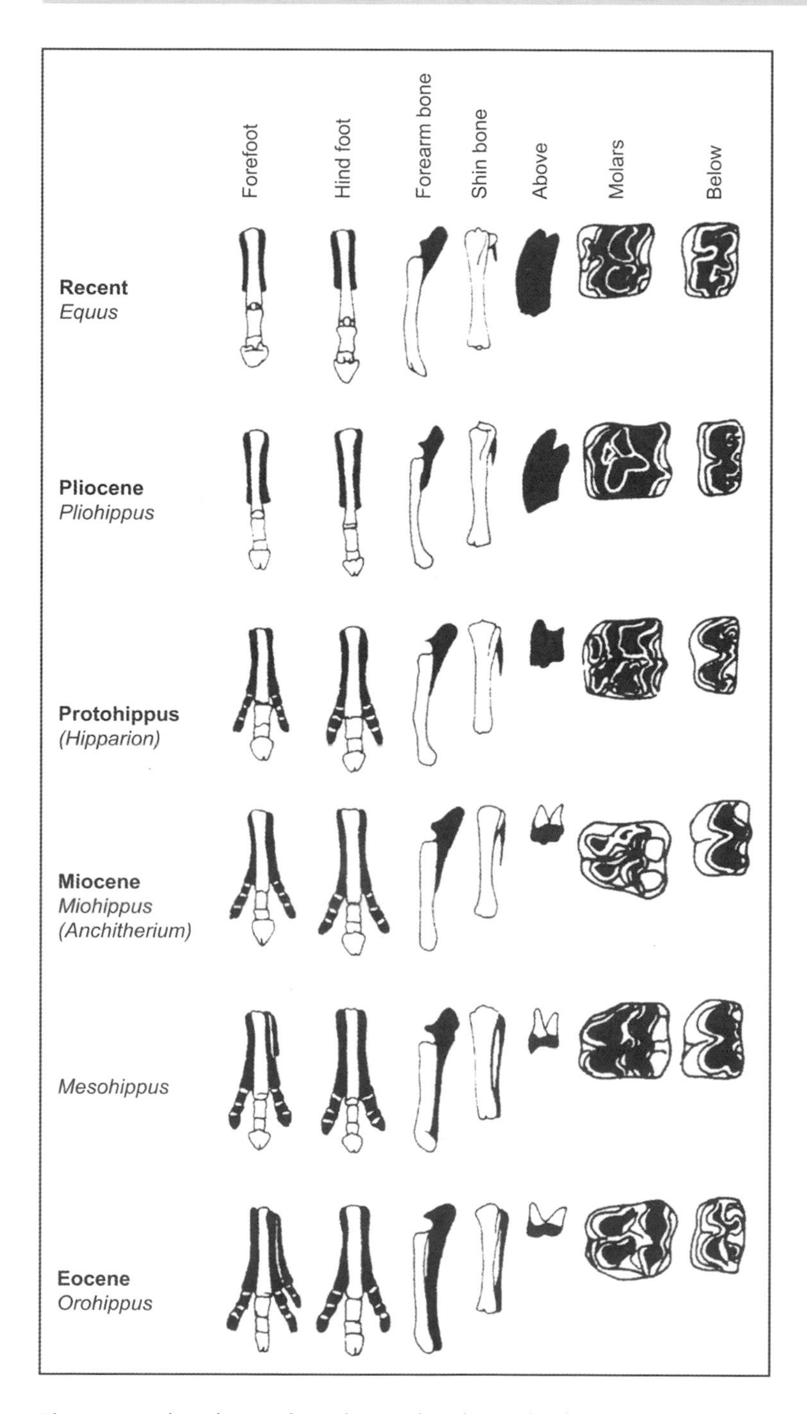

Figure 7.3. This chart is based on Othniel Marsh's horse progression from 1879 (still without *Eohippus*). Here, reduced bones, dentin, and dental cement are shown in black. The crowns of the maxillary molars, seen lengthwise (in front view), are silhouettes.

therium. The fragments fit comfortably inside a cigar box. Some sites in these godforsaken lands were rich in fossils, others were barren. Sharp vision, insatiable curiosity, an instinct for where to search, boldness, physical stamina, and not least of all a steady dose of good luck were required in order to return to civilization with full bags.

It would be too much to describe in detail all the expeditions that followed. Not to be overlooked, however, is the American Museum of Natural History's 1910 expedition, under the leadership of William Sinclair and Walter Granger, into the central Bighorn Basin. It still ranks as one of the most successful undertakings ever in this area. In addition to acquiring numerous fossil discoveries, Sinclair and Granger succeeded in distinguishing different faunas. And

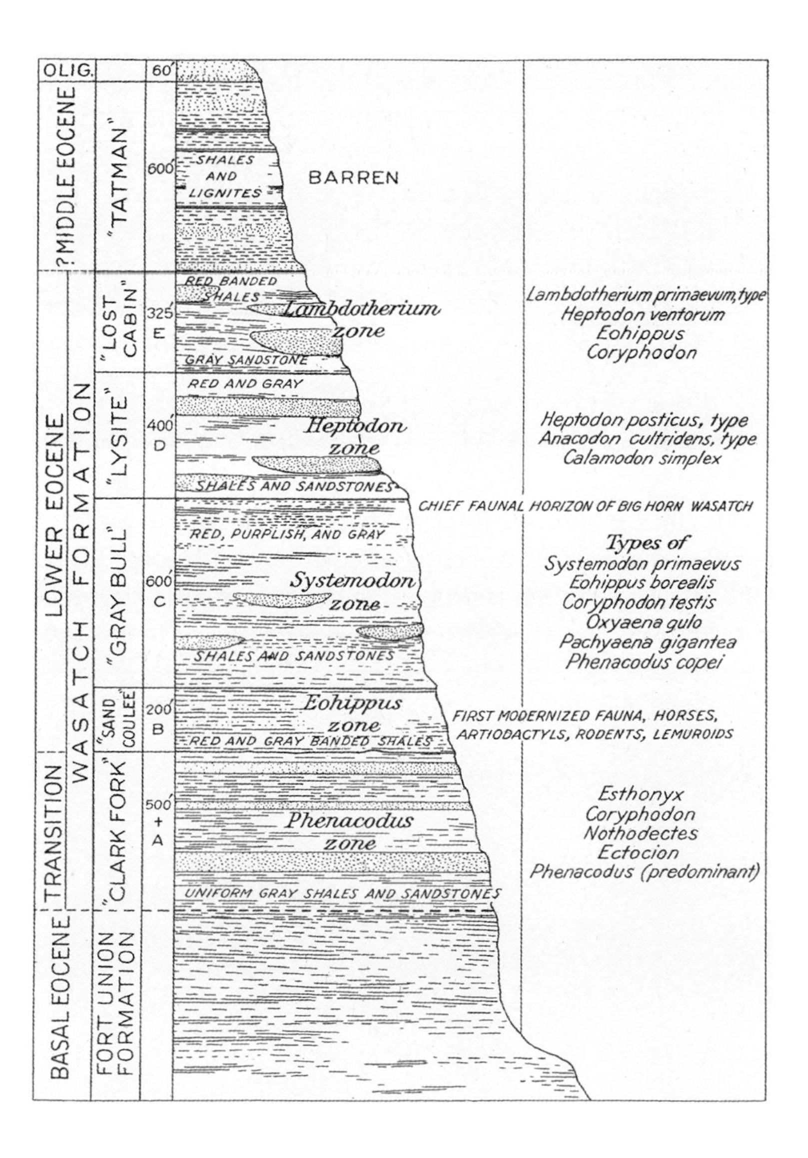

Figure 7.6. Sequence of Paleocene (basal Eocene + transition) and Eocene stratigraphic layers with pertinent fossil mammals in the Clarks Fork and Bighorn Basin. Walter Granger compiled this sequence in 1918.

they recognized that these emerged from different stratigraphic layers (Figure 7.6). For the first time, an idea arose about the stratigraphic sequence of the Early Eocene, the so-called Wasatchian, named after a Native American tribe that lived in the area. With the chronology of the 1,500-meter-thick sequence of layers, one could trace the development of many species over a several-million-year interval.

Figure 7.7. Walter Granger astride his horse Mormon at a field site in the Eocene of Utah in 1895.

The horse fossils that the expedition collected were sent to the East Coast. There, the millionaire Henry Fairfield Osborn (1857–1935), his student and assistant the genial William Diller Matthew (1871–1930), and that connoisseur of fossil horses Walter Granger (1872–1941) worked at the American Museum of Natural History in New York. (Osborn was renowned for his weighty volumes on the evolution of proboscideans [Proboscidea], titanotheres [known today as brontotheres], and horses.) But the men did not limit themselves to work in their labs and collections; they also participated from time to time in western expeditions (Figure 7.7).

When one thinks of fossil horses, one normally thinks of teeth and bones. Tilly Edinger (1897–1967) demonstrated that it is possible to reconstruct the evolution of sensitive structures, such as brains, at least in broad outlines. Tilly was the daughter of Ludwig Edinger (1855–1918), a famous neurologist from Frankfurt am Main. She worked at the Senckenberg Museum in Frankfurt, until she had to quit because of the increasingly dangerous anti-Semitic Nazi sentiments. She fled Germany in 1939, going to the United States via England. She continued her pioneering studies after she arrived in the United States.

How is it possible to study structures as delicate as brains using fossils? The answer is found in the biomechanical conditions of ontogenetic cranial development, which means the development of the skull from the embryo to the adult. The shape of the cranium is determined by external forces, like muscle traction and gravity, and by the growth of organs inside the skull, such as the brain, the eyes, and the dentition. Through its hydraulic construction and development, the brain has a tendency to expand spherically in all directions. Consequently, throughout its development it presses on the surrounding bony braincase, which is developing at the same time. In this way, the outside structures and the windings of the brain are copied on the interior of the braincase. It is possible with fossils to obtain interior casts of the brain case. These internal casts, known as endocranial casts or endocasts, are extracted with the help of latex and allow for three-dimensional reconstructions of the brain. Some brain cases of fossil horse skulls survived the fossilization process as a cavity, which allows for molding. Sometimes this process even happens in a natural way, as when circulating water precipitates calcium deposits inside of the hollows of the skull. In 1948, after intensive and detailed studies, Tilly Edinger presented a progression of endocasts of fossil horses, documenting the evolutionary development of external brain anatomy from *Eohippus* (= *Hyracotherium* or *Protorohippus*) via *Mesohippus, Merychippus,* and *Pliohippus,* right up to *Equus* (Figure 7.8). The most important developments were evidenced by the increasing dominance of the brain, particularly a disproportionate expansion of the neocortex. This type of brain, which characterizes the modern ungulates, did not develop prior to the Late Miocene. Later the North American mammal paleontologist Leonard Radinsky (1937–1985) showed that Tilly Edinger had mistaken the reported *Eohippus* endocast for that of a condylarth and that the brain of a true *Hyracotherium* had clearly already developed to a greater degree. In no way, however, does that undermine the groundbreaking and pioneering work of Tilly Edinger. Radinsky also showed that the expanding proportion of the frontal lobes was probably accompanied by increasing tactility of the lips. This is pre-

Figure 7.8. Development of the horse brain over 55 million years, from *Hyracotherium* to *Equus*.

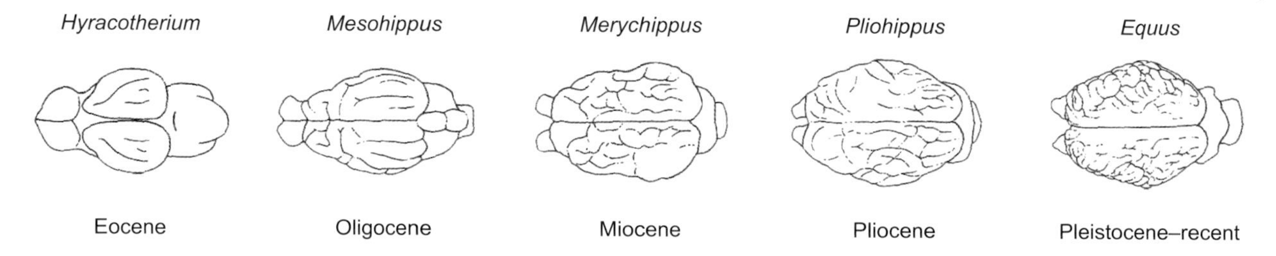

sumably as important as the olfactory sense is to herbivores that eat without being able to see their food.

Besides the expansion of the horses' phylogenetic tree through numerous fossil discoveries, paleobiological analyses revealed additional interesting aspects. This field of research, which Kowalevsky had introduced, is still fruitful. Matthew, Osborn's assistant, came to the conclusion in 1926 that the first horses were not omnivores, as Kowalevsky had presumed 50 years before, but browsers. Matthew based his hypothesis on comparisons with similarly built cheek teeth of modern mammals, like the South American pudu and brocket deer of the Old World. I myself had the pleasure of being able to show Matthew's daughter at the Senckenberg Museum many years later that this was the case. The first analyses of the intestinal contents of the Messel horses clearly demonstrated that these Eocene horses had fed mainly on leaves, and hence that her father had once drawn the right conclusion.

So it seems that Matthew's revision of Kowalevsky's hypothesis was valid. This was not the case with his explanation of the development of the locomotory apparatus. Kowalevsky explained this in terms of an improvement, an economization of the construction. Matthew believed that the changes seen in the extremities and the teeth were simply the product of environmental changes. In 1926, he wrote with reference to *Eohippus*: "It was probably a forest dweller, browsing on leaves and succulent herbage the flexible four-toed foot adapted to the irregularities of the woods, to quick shifts and dodging behind cover, but not to speed over a uniform open surface." With regard to further development, he continued: "The change in limbs and feet is . . . equally explained as an adaptation to the progressively dry climate and open plains environment." This apodictic-sounding statement is even more astonishing when one realizes that Matthew once described the development of the extremities as an improvement in their construction, especially in connection with increasing body size. Matthew failed to mention Kowalevsky's ideas either because he was unfamiliar with the appropriate publication—it was written in German—or he assumed Kowalevsky's monograph on the genus *Anthracotherium* provided no information on horse evolution. *Anthracotherium* was well known as an artiodactyl at that time. Nevertheless, from that moment on, the phylogenetic development of the horses has uniformly been depicted as an adaptation to a change in the natural environment—from original browsing jungle inhabitants to grazing steppe and savannah animals. This was done in clear contradiction to evidence, because the reduction from the original five hooves per leg demonstrably began more than 30 million years before the environmental change. In addition, with the transition from predominantly jungles to steppes and savannahs the development to equids was in no way complete. Indeed one-hoofed equids first emerged much later than the environmental change, only approximately 4 to 5 million years ago. Thus in this case it is evident that there is absolutely no connection between environmental changes and the development of structures.

A prerequisite for paleobiological studies is an understanding of corresponding biological, and hence observable, processes and conditions. Locomotor studies of modern horses illustrate this idea. Four legs permit a variety of locomotory patterns. Horse connoisseurs speak about no fewer than 12 different gaits. However, they are all variations on only four base rhythms—the walk, the trot, the canter, and the gallop. It is amazing how little was known about these modes of locomotion for such a long time. Horses are simply too fast for human eyes and with their four legs, too diverse in their locomotion. The lack of understanding of the locomotory patterns is clearly expressed in early drawings and paintings of galloping horses (Figure 7.9). Artists depicted horses with wide, forward, outstretched forelegs and likewise wide, backward, outstretched hind legs. Such a position is not only unnatural but also practically impossible for a horse. It was a photographer, Eadweard Muybridge, who in 1872 on a farm in California began to find out what really occurs when a horse runs. With his custom-made camera equipment developed for that purpose, he was able to record the exact sequences of the gaits of moving horses (Figure 7.10). In 1884 and 1885, Muybridge continued his studies, which he extended to other animals and finally even humans, at the University of Pennsylvania. For these studies, he used 24 single-frame cameras and 2 large cameras for over-

Figure 7.9. A horse race at Newmarket (England) in 1799. Note the wide, outstretched fore- and hind legs of the horses. Today we know that such a gallop is biomechanically impossible, but the human eyes and the brain are not capable of processing the detailed motion sequence. The picture is on exhibit in the Deutsches Pferdemuseum in Verden.

Figure 7.10. Muybridge's photographic sequence of a galloping horse.

views. The apertures of the single-frame cameras were triggered by tripwires, which the horses touched one after the other. In this way, Muybridge succeeded for the first time, in documenting the different gaits in a chronological succession of single phases.

These picture sequences—which today could be easily obtained with a video camera—also enabled for the first time the reconstruction of the locomotory patterns of fossil horses with the aid of skeletal finds. Modern computer animation even permits the filmlike re-creation of locomotion of long-dead horses. Such computer animation of the Messel primitive horse *Eurohippus messelensis* was developed at the Institute of Systematic Zoology and Evolutionary Biology in Jena—the former institute of Ernst Haeckel—under the direction of Martin Fischer. Another version is under way at the Forschungsinstitut Senckenberg in Frankfurt, which is collaborating with the Hessian Broadcasting System (Hessischer Rundfunk). The next step would be the virtual reanimation of 47-million-year-old horses in their natural environment.

Fossil horses have been found on all continents, with the exception of Australia and Antarctica (see Chapter 8). With all of these discoveries, it has become clear that in considering the evolution of horses we are not dealing with a linear, orthogenetic, or orthoselective development, like Marsh and others had presumed. What we are dealing with is a rather complex system of developmental branching lineages in completely different directions and detailed distributions, as suggested in G. G. Simpson's phylogenetic tree of 1951 (Figure 7.11). For example, alongside the development of high-crowned molars and one-hoofed fossil horses, there is the retention of low-crowned dentitions and three-hoofed limbs and the evolution of massive size in species such as *Hypohippus* and *Megahippus*. Apparently they did not venture out of the jungle to the increasingly expanding steppes and savannahs. On the other hand, reversions back to the forest and to leafy food occurred despite the retention of high-crowned molars, as demonstrated by the European species *Hippotherium primigenium* (see Figure 6.2). Additionally, alongside general phylogenetic size increases dwarfing also occurred, as with North American *Nannippus* and within the Eurasian genus *Hipparion* (*H. periafricanum*). Such developments have yet to be satisfactorily explained in the evolutionary history of the horse. Finally pseudo horses appear, like the European palaeotheres or the South American *Thoatherium*, which have nothing to do with the phylogeny of horses, but rather illustrate convergent or parallel developments (see Chapter 9). Evolution proceeds not dogmatically, but opportunistically. The only routing criterion for its reconstruction, besides the timeline, is the improvement of energy balance (see Chapter 6).

The investigation of the phylogenetic development of horses, however, progressed through new finds, new hypotheses, and new paleobiological analyses. Important new insight is also owed to the technological advancements of research methods. This applies of course to the scanning electron microscope (SEM), through which the unknown was made visible—like intestinal contents or soft body contour through bacteria (bacteriography). By comparing findings with modern ungulates, structural investigations of occlusal patterns and wear surfaces (macrowear, mesowear, microwear) permit detailed statements about food and nutrition (paleodietary analyses). The advancement of imaging methods (x-ray and CT) has also provided insights into hitherto unknown worlds. Biological and geo-

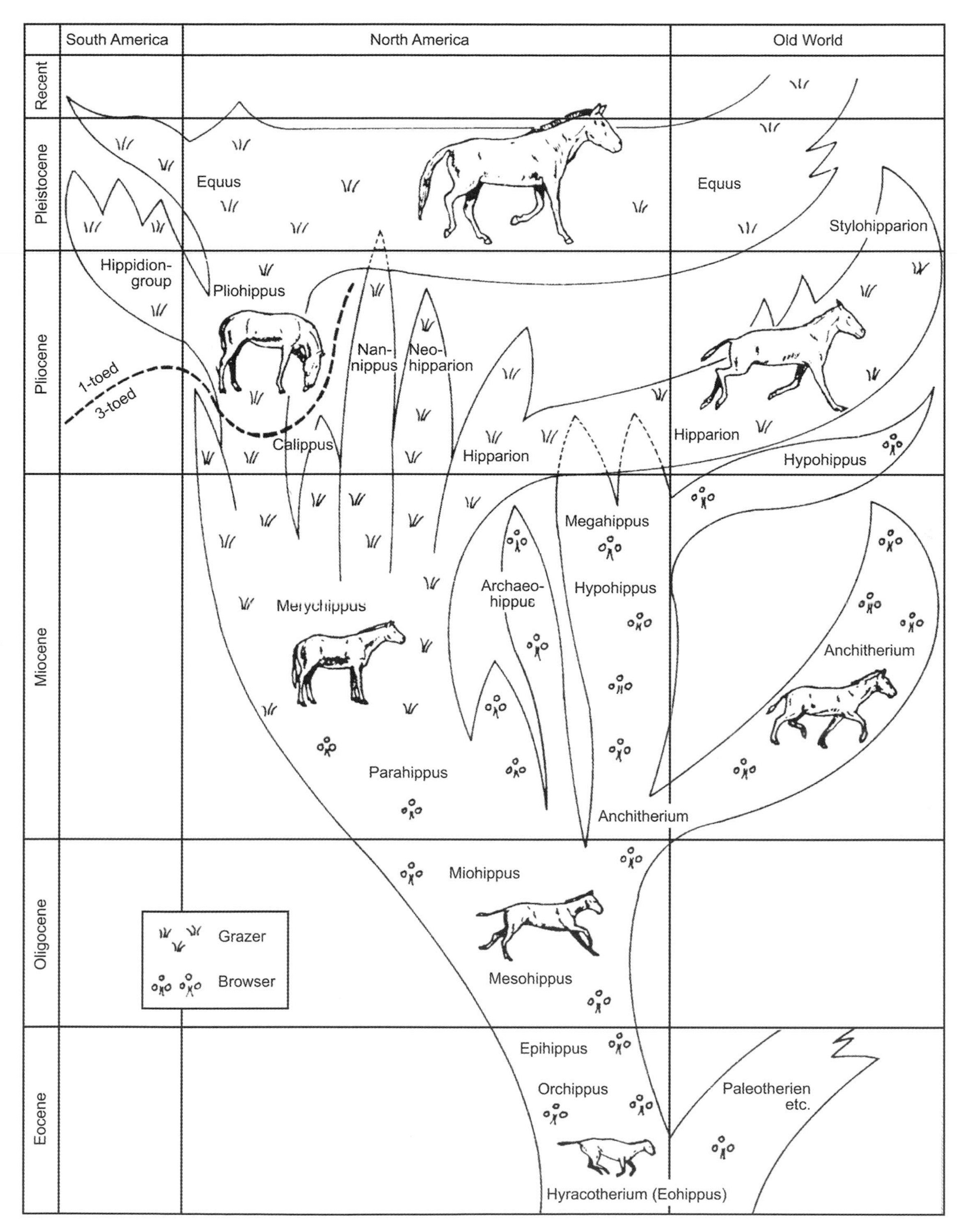

Figure 7.11. G. G. Simpson's horse phylogeny from 1951, which hints at the diversity of developmental lineages. Two years later Simpson would acknowledge the chart's oversimplification.

chemical methods also play an increasingly important role. Analyses of naturally occurring elements in teeth and bones, such as strontium or carbon isotopes, allow conclusions about diet, sometimes even providing information about the area where the respective animals lived. That in turn enables us to make inferences on the lifestyle and the ecological demands of those animals, on transport and accumulation processes, and finally on the formation of fossil lagerstätten. Chemical changes in the fossils and their environment point to fossil preservation processes. Analyses of embedded sediments allow the reconstruction of climate changes in the geological past. Finally, molecular genetics have shed light on the classification and phylogeny of horses, at least during the ice ages.

Many of these approaches not only allow us to acquire knowledge but also have practical implications for humans. One example is the study of the formation of natural resource deposits. But science is indivisible. One cannot say *this* scientific effort is worthwhile because it directly addresses some human need, but *that* scientific effort is not worthwhile because it has no immediately discernible benefit for humans. All knowledge that humans obtain about themselves and their environment is valuable. It allows us to improve our and other species' chances of survival and to aid the further development of humankind. One never knows in advance what new insights will someday lead to practical applications. Many important innovations have come from basic research. Mankind's innate curiosity and our ability to imagine the unimaginable serve us well in the pursuit of knowledge. Ultimately, like art, pure science is part of our culture.

And like art, science need not always be unprofitable. Think of the increasing public attention on fossil finds and evolutionary questions. Almost all natural history museums have recorded increases in visitor attendance. Films like *Jurassic Park* have been true blockbusters. Together with media reports and book publications, such endeavors can mean billions of dollars in sales figures worldwide. Sciences like paleontology have no reason to shy away from business aspects. But profits are not the only rationale for the existence of the geosciences. We people live on the earth, and we live from the earth.

CHAPTER EIGHT

Evolution and Expansion of the Horses

UP TO NOW WE HAVE LOOKED at the evolution of the horses considering their construction, their way of life, and their environment. Now we want to ask if there were geological factors that influenced the evolution of the horses. Candidates could be orogenesis (mountain building), continental drift (plate tectonics), climatic development, as well as isolation due to epicontinental flooding (transgressions). Climatic development is discussed in Chapter 6. Orogenesis plays a particularly important role in climatic development with respect to air mass circulation, temperature, and water content. Additionally, orogenesis and the development of rift valleys may lead to the isolation of populations and their subsequent endemic evolution. For example, a separate development occurred when the orogenesis of the Pyrenees led to a temporary isolation of the Iberian Peninsula from the European continent during the Eocene. Isolation as a result of marine transgressions also played a role in the endemic development of the mammal fauna on the Cantabrian Island during the Late Eocene. The palaeotheres, the relatives of the dawn horses of Eocene times, showed strange developments such as upper molars with extremely high inward-bending external walls. The fragmentation of the European continent during the Eocene leading to an island archipelago also may have played a role in the splitting of the horses into numerous independent phylogenetic lineages. Plate tectonics were also extremely important, which becomes evident with the beginning of the fossil record of the horse's evolutionary development at the start of the Eocene 55 million years ago. At this time the first horses of the *Hyracotherium* type appear contemporaneously on all northern continents, with the possible exception of Asia. This fact is only understandable when we consider intercontinental land bridges like the one over the North Atlantic (although this later broke down due to plate tectonics, rising sea levels, shifting climatic zones, or a combination of all three factors). In any case, there are no Paleocene ancestors of horses known from any northern continent. Therefore, the earliest horses of the fossil record must have come together with some other perissodactyls, artiodactyls, chiropterans (bats), and the primates, from one of the southern continents—South America, Africa, or India. All of these mammals appear at the beginning of the Eocene as part of an immigration wave on the northern continents. Where did all these immigrants come from?

The fossil record confirms that South America was more or less isolated until the end of the Tertiary. At the time in question it was not inhabited by ungulates, bats, or primates. Consequently, the immigration wave could not have come from there. The situation in Africa and India, the only other southern continents relevant in this context, is different. At that time, Africa was in close contact with the Iberian Peninsula and with Europe via the Strait of Gibraltar. At about the same time, the Indian subcontinent collided with the original Asiatic continent after a long period of northward drift (Figure 8.1). However, direct ancestors of horses are still unknown from these continents. Fossil localities that could deliver relevant witnesses of the Early Eocene or the Late Paleocene have only recently become known from Africa and India. Therefore, we have only circumstantial evidence to rely on. The Hyracoidea, or dassies, which are relatives of horses, lend support to an African origin of horses. Up to the Late Miocene, dassies are known exclusively from Africa, including North Africa. If horses are closely related to hyracoids, their common ascendants must have come from Africa. Another indication for an African origin of horses is the fact that the earliest record of the Euprimates, *Altiatlasius,* was found in Late Paleocene deposits in Morocco. If horses entered the northern continents with the Euprimates, this would point to an African origin of the immigration wave. On the other hand, the systematic position and phylogenetic relationships of *Tanzanycteris mannardi* (the oldest bat from Africa south of the Sahara) from the Eocene crater lake of Mahenge (Tanzania) are uncertain. A somewhat limited faunal exchange between Africa and Europe is, however, indicated by the anteater *Eurotamandua joresi,* the ostrich *Palaeotis weigelti,* the phorusrhacid (a now extinct bird family) *Aenigmavis sapaea,* and the crocodile *Bergisuchus dietrichbergi,* all of which occur at Messel. They would have

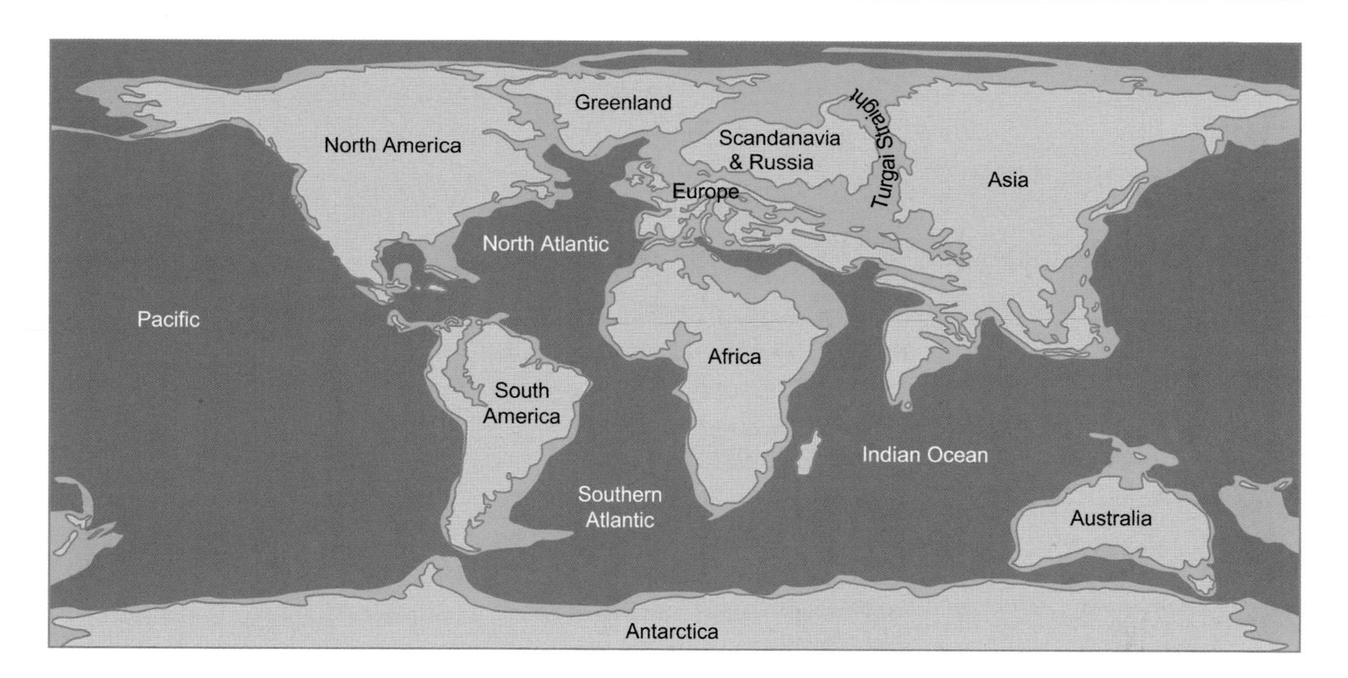

Figure 8.1. Location and contours of the continents during the Eocene (around 50 million years ago). Light blue = flooded continental areas.

come to Europe by way of Africa. But except for the dassies, and in spite of rich fossil localities, no relatives of dawn horses have yet been discovered up to now from North Africa. Therefore all of our hope depends on the huge area of sub-Saharan Africa, which has not yet delivered any evidence of fossil mammals that could serve as direct proof for the relevant time interval.

On the other hand, horses are much more closely related to tapirs than to hyracoids. Tapirs apparently stem from Asia, where they dominated the Eocene faunas with a multitude of genera and species. It is also from Asia that *Radinskya yupingae* was found, a skull from the Late Paleocene of China, which shows similarities with *Hyracotherium* (Figure 8.2). The genus *Radinskya*, which resembles early tapirs more than *Hyracotherium* particularly with respect to its dentition, is nowadays classified with the Altungulata, a super order that comprises the proboscideans (Proboscidea), the sea cows (Sirenia), the dassies (Hyracoidea), and the odd-toed ungulates (Perissodactyla). If there is a phylogenetic connection between tapirs and horses, the morphological evidence and intermediate fossils are lacking. With respect to the horses, *Radinskya* could also represent a very distant relative, or it could be the result of parallel or even convergent development. Therefore the origin of *Hyracotherium*, and with that of the whole phylogenetic tree of the horses, remains in the dark. The genuine ancestors of the horses appear on the northern continents no earlier than the beginning of the Eocene, but where exactly did they first turn up?

An answer to this question could possibly be provided by a new discovery. Only recently, the boundary between the Paleocene and the Eocene was recognized to be characterized by a worldwide event (see Chapter 3). For a geologically short time interval of about 20,000 years, a considerable warming of the earth's atmosphere occurred due to a sudden increase in the greenhouse gas carbon dioxide. This increase probably is tied to submarine outbursts of hydrate concentrations, which influenced life on land and in the sea. Evidence is present in deep-sea sediments as well as on the continents. Being a geologically

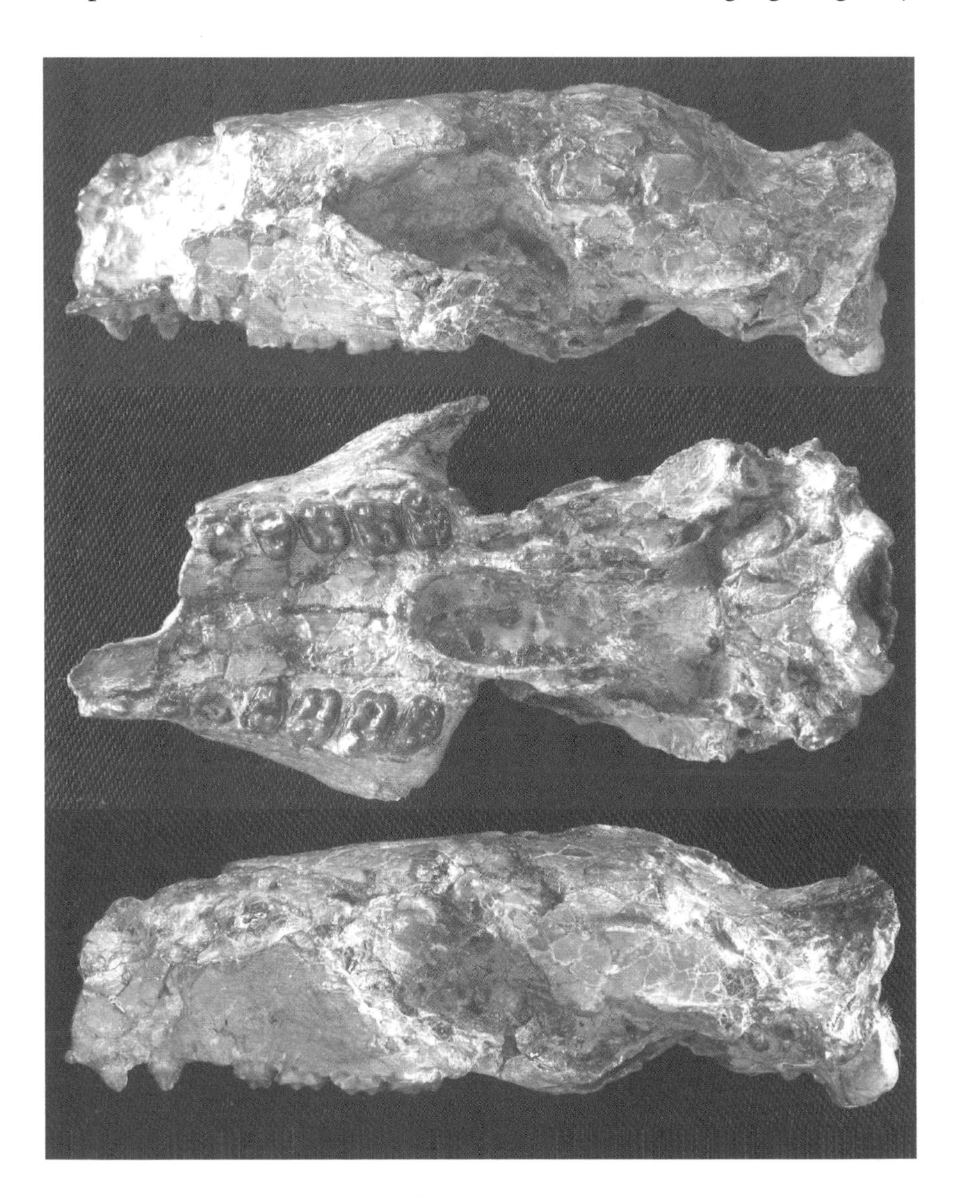

Figure 8.2. Skull (length about 9 cm) of *Radinskya yupingae* from the Late Paleocene of China (*top to bottom*): seen from the left side; from below; from the right side. The modest molarization of the premolars (anterior cheek teeth) is similar to that of the earliest dawn horses; the emphasized transversal ridges of the molars are more representative of the tapiroids.

short time event, this PETM (Paleocene-Eocene-Thermal-Maximum) offers the possibility of tracing the Paleocene-Eocene boundary worldwide and correlating the sediments in its realm. This in turn offers the potential for analyzing when and where the first equids appeared on the northern continents, thus ultimately providing evidence on their origin. However, up to now the PETM has been found with fossil equids only in North America, specifically in Wyoming. In China, dawn horses of *Hyracotherium* type are lacking, while in Europe the PETM has yet to be correlated with findings of dawn horses. Therefore, for the moment we must wait and see. Because it is not yet possible to tell the global story, I will describe the evolutionary history of the horses on a continent-by-continent basis.

North America

There is no doubt that since the Eocene horses existed in North America. It was there that most of their evolution actually took place. From time to time, branches ventured into the Old World and also into South America at the end of the Tertiary (Figure 8.3a–d). Where the horses came from and which path they took when they first arrived in North America is still unknown. In spite of some similarity, no continuous development was recognized from primeval ungulates (Condylarthra), and more particularly from the phenacodonts. Equids could not have come from South or Central America or from the Arctic because no candidates for ancestors lived in any of those places. Therefore, only two possibilities remain: They could have come from Asia by way of the Bering land bridge, which at that time existed between Alaska and Siberia, or they could have come from Europe by way of the so-called Thule land bridge, which existed over England, the Faroe and Shetland Islands, Iceland, and Greenland. In favor of a European immigration is the presence of the most primitive *Hyracotherium* in the Early Eocene (Late Paleocene?) of Silveirinha (Portugal). Also supporting this route is the fact that up to now there have been no representatives or close relatives of *Hyracotherium* known from the Paleocene-Eocene boundary in Asia. And finally, the fossil record shows that the dawn horse skeletons from the early Middle Eocene of Messel are more primitive and have less specialized foot construction than the considerably older hyracotheres from the Early Eocene of North America.

Advanced specializations for running distinguish the North American hyracotheres—which have been differentiated into *Hyracotherium, Sifrhippus, Minippus, Arenahippus, Xenicohippus, Protorohippus,* and *Eohippus*—from their younger relatives from Messel. In

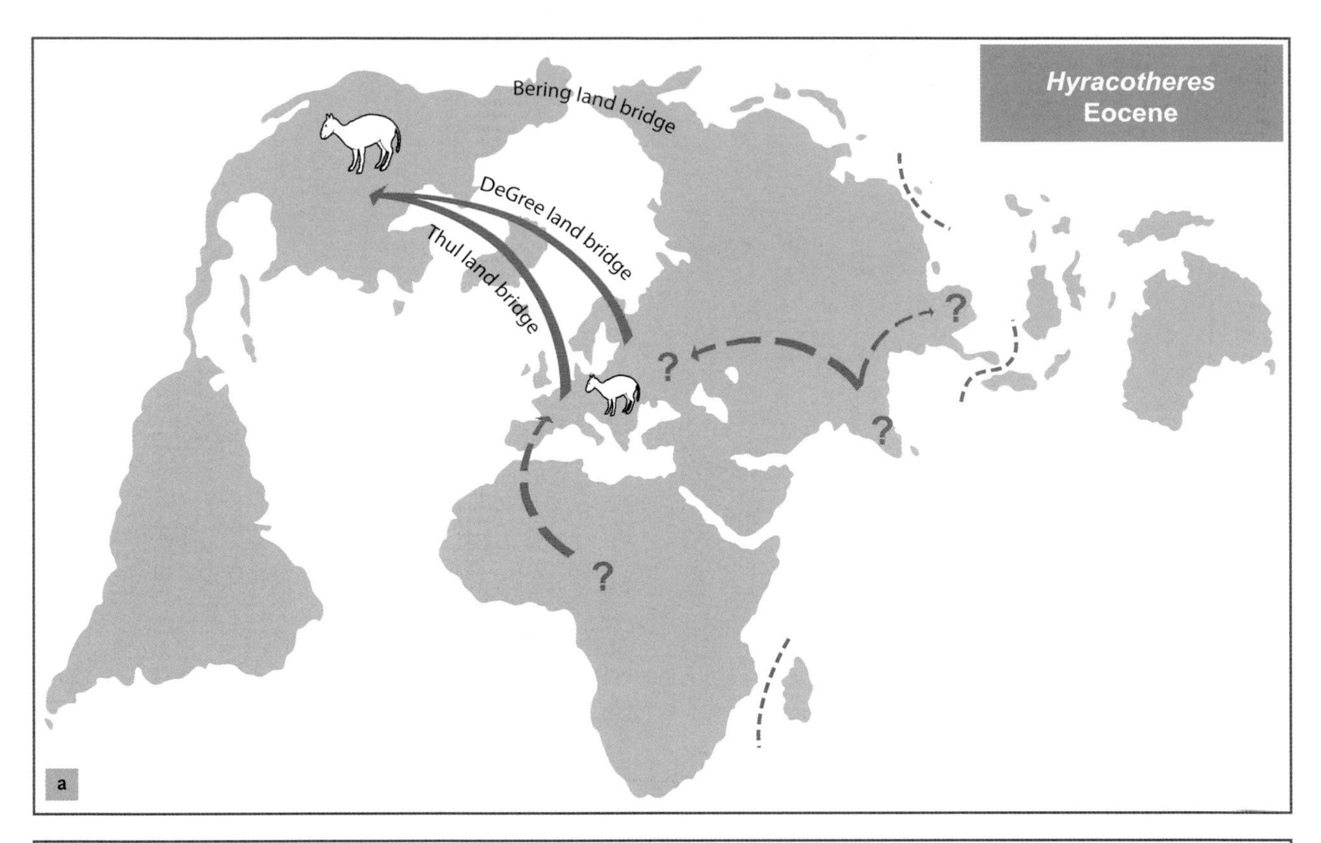

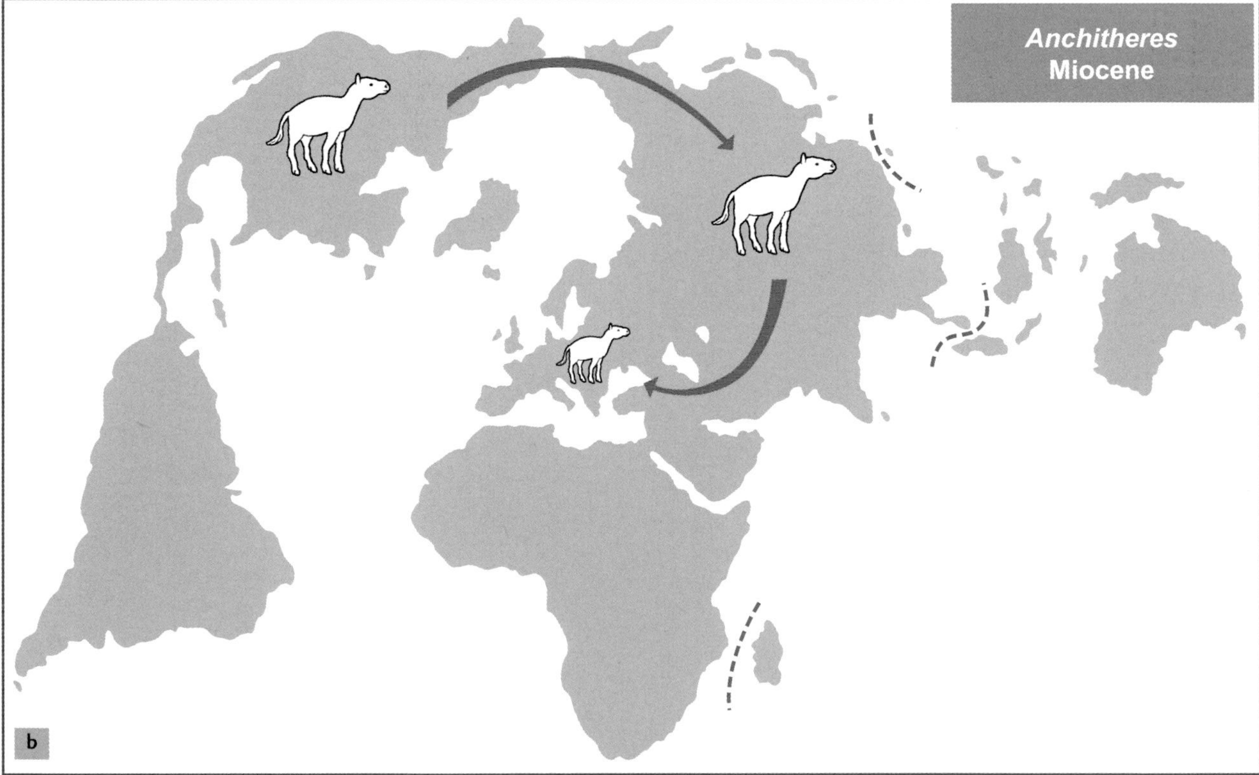

Figure 8.3. Origin and expansion of horses in four phases: (a) the hyracotheres (Eocene), (b) the anchitheres (Miocene), (c) the hipparions (Late Miocene–Pliocene; in East Africa until the Middle Pleistocene), (d) *Equus* (Pliocene–Pleistocene, except for return to North America aboard Spanish caravels during the sixteenth century).

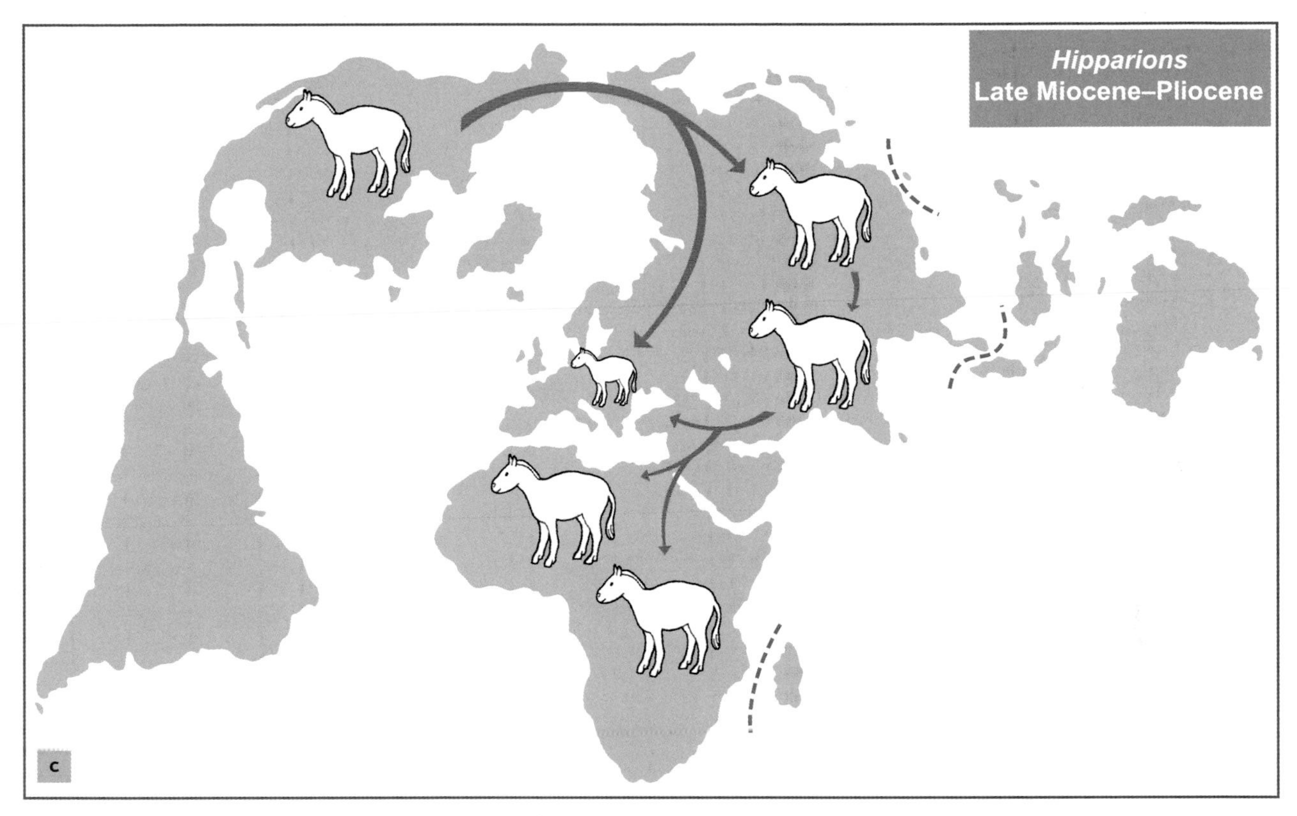
Hipparions
Late Miocene–Pliocene
c

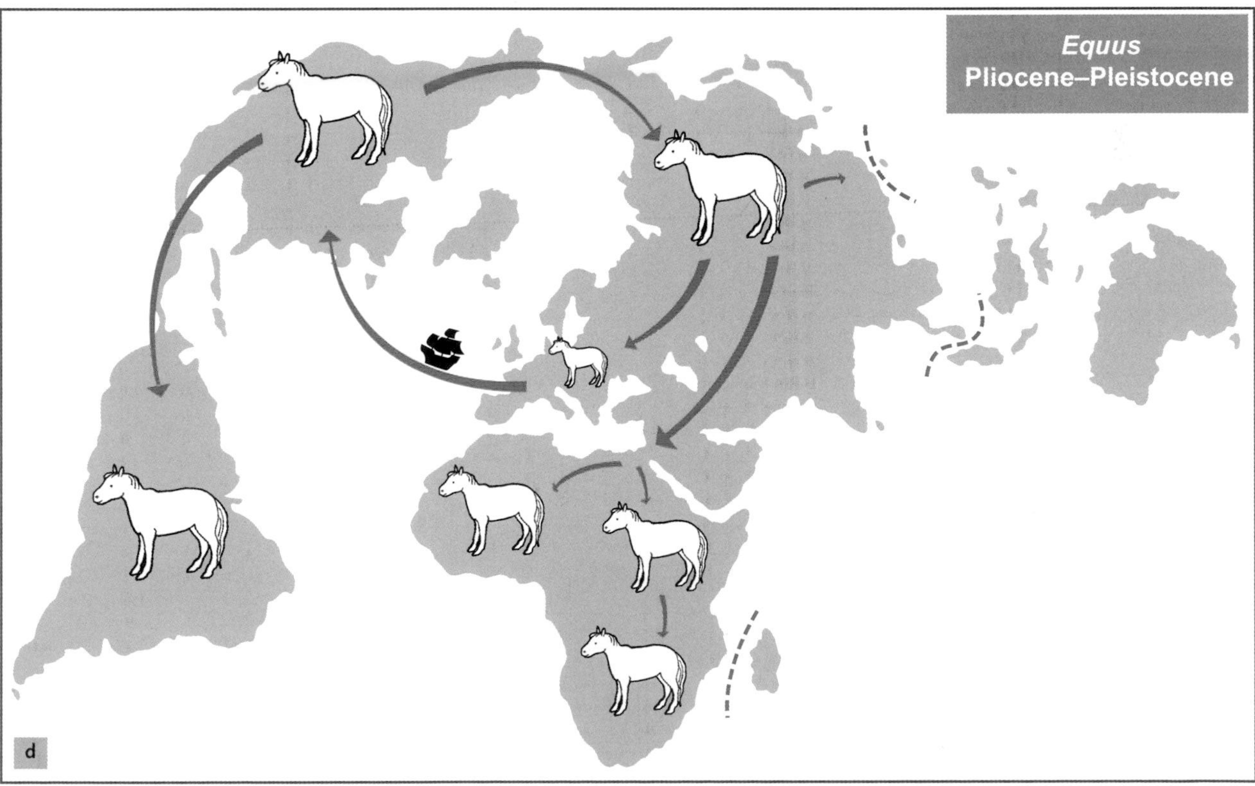
Equus
Pliocene–Pleistocene
d

North American hyracotheres, the limbs and the feet are already considerably elongated—hence more cursorially specialized—when compared with the dawn horses from Messel. While *Eurohippus* from Messel still displays the remnants of the first digit in the forefeet and the fifth digit in the hind feet, these are already completely reduced in the North American hyracotheres (see Chapter 5).

Consequently, European *Hyracotherium*—a genus that is more or less the same age as the North American hyracotheres—should have been less advanced than the Messel horses in this respect. Unfortunately, no postcranial bones have been found for European *Hyracotherium*. In any case, based on the available evidence, the development and also the migration of the hyracotheres could only have occurred from Europe to North America and not vice versa. Another argument supporting this idea is that the dentally most primitive Eocene horse, *Hallensia*, is only known from Europe. As the migration of the earliest Eocene primate, *Teilhardina*, from Asia to North America via Europe proves, such an intercontinental dispersal could take place within an interval of less than 25,000 years.

Starting from *Hyracotherium (Sifrhippus) sandrae*—the earliest North American hyracothere—equids evolved on this continent via *Orohippus* (Middle Eocene) and *Epihippus* (Late Eocene) into *Mesohippus* (Early Oligocene) and *Miohippus* (Late Oligocene) (see Figures 6.1, 7.3, and 7.11). This development was characterized by a continuous molarization of the premolars and a reduction of the lateral toes. During this process the cheek teeth changed from bunodont to lophodont—hence from a bumpy occlusal pattern to occlusal ridges—but remained low crowned. As analyses of the gut contents of the equids from Messel have shown, such an occlusal pattern seems to be typical for browsers, which could also occasionally have fed on fruits (see Chapter 5).

During the Early Miocene—around 20 million years ago—the development of the horses split into several lineages (see Figure 11.1) forming two larger groups. One originated in the Early Miocene from the genus *Parahippus* and developed into the Middle Miocene *Merychippus*, which were the first horses that fed on grass (graminivorous). *Merychippus* is characterized by increasing high-crowned (hypsodont) cheek teeth with a very complex chewing surface (occlusal pattern). The other group that developed during the Early Miocene originated with the genus *Kalobatippus*. These dawn horses were still browsers but had increased considerably in size. Within this group, genera such as *Anchitherium* developed. Also stemming from this group was *Hypohippus*, which would later develop during the Late Miocene into *Megahippus*, reaching a size similar to modern horses.

The group of graminivorous horses split toward the end of the Miocene—12.5 million years ago—into a large group of genera that included *Protohippus, Calippus, Astrohippus, Pliohippus, Dinohippus, Pseudohipparion, Neohipparion, Nannippus,* and *Cormohipparion.* At least three times, in parallel, all side toes were completely reduced within the genera *Astrohippus, Pliohippus,* and *Dinohippus.* Thus monodactyls developed. Other features—especially the reduction of facial fossae in front of the eyes—demonstrate that it was in this group that around 3.9 million years ago, on the threshold of the ice age, the monodactyl genus *Equus* developed. This genus is represented today in the form of the modern horses (*Equus caballus*), asses (*Equus asinus*), half-asses (*Equus hemionus*), and zebras (*Equus zebra, E. burchelli,* and *E. grevyi*). The quagga (*Equus quagga*) was hunted to extinction toward the end of the nineteenth century. All other genera and species of horses died out before, during, or just after the ice age.

The question of how the genus *Equus* developed during the early Pleistocene, around 1 to 2 million years ago, is problematic. The first record of the genus was a single skull found in the Anza Borrego Desert of Southern California. It was erroneously described under the name *Equus simplicidens*; this name had already been applied to finds from 3.5-million-year-old layers of the famous Hagerman Quarry in Idaho. Those finds, however, were placed meanwhile in the genus *Plesippus* and *shoshonensis* was substituted for the species name *simplicidens.* Be that as it may, it is still open as to how *Equus "simplicidens"* developed into *Equus scotti* and *E. excelsus* of the Middle Pleistocene of North America. Both species are the source of the last genuine North American horses, which survived into postglacial times, around 8,000 years ago, in the forms of *E. lambei* and *E. occidentalis.*

Another group of horses in the Pleistocene is known only from North America. These are the New World stilt-legged (NWSL) horses. As the name indicates, these are horses with exceptionally long, slender built legs. Limited to Central and North America, from Mexico up to Alaska, the NWSL never reached other continents. In their limb proportions they resembled the Asian half-asses of the *Hemionus* group. Evidently, they represented a North American equivalent without a direct relationship.

During the Miocene, lineages of North American horses repeatedly branched off and ventured into the Old World by way of a land bridge where the Bering Strait now separates Alaska from Siberia. A first wave occurred at the beginning of the Miocene with the genus *Anchitherium,* browsing horses with relatively low-crowned lophodont dentition. This genus developed from *Kalobatippus* into the

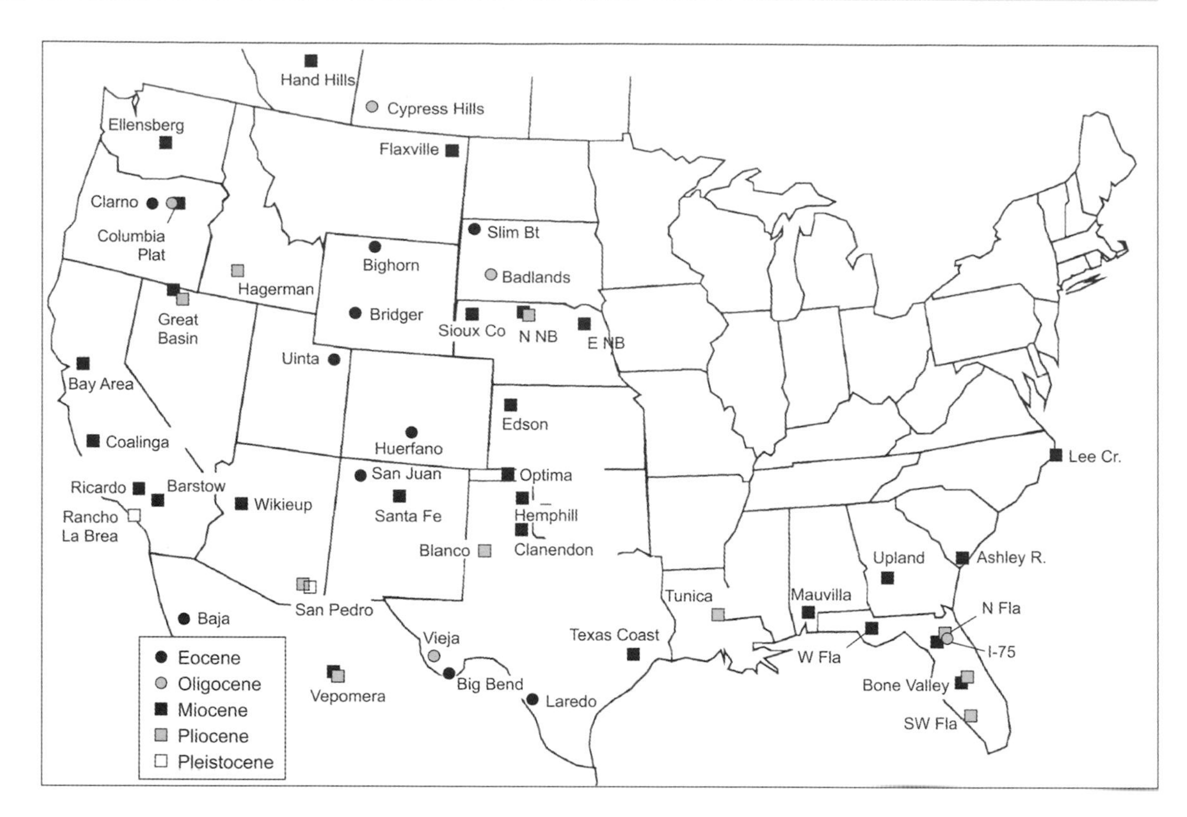

Figure 8.4. Select fossil horse sites in North America. Of the many Pleistocene sites only Rancho La Brea and San Pedro are shown.

large *Megahippus* and *Hypohippus*. It later expanded into Europe (Figure 8.3b). The giant species *Sinohippus* also derived from *Anchitherium*. It occurred no earlier than the Middle Miocene and remained restricted to China. It is remarkable that in spite of their very large size *Megahippus* and *Sinohippus* remained browsers. Toward the Late Miocene, the grass-eating North American horses twice sent descendants of *Cormohipparion* into the Old World via the Bering Strait throughout Asia as far as Europe (Figure 8.3c). The genus *Hippotherium* appeared around 11 million years ago. It remained restricted to the northern parts of Asia. Recent investigations have shown that these horses returned at least partially to a browsing lifestyle, although they retained the high-crowned cheek teeth of their ancestors (see Figure 6.2). The fossil record from Germany and Spain shows that *Hippotherium* met the last survivors of the genus *Anchitherium* in Europe before their extinction (which soon followed). Somewhat later the first true Hipparions, which besides *Hipparion* comprised genera such as *Cremohipparion*, *Sivalhippus*, and

Proboscidipparion, appeared in the Old World. *Hipparion* was the first horse to enter Africa, setting hoof on the continent some 10 million years ago. This branch survived into the middle of the ice age as *Stylohipparion* (Figure 8.3c).

Branches of the North American horses not only ventured into the Old World but also into Central and South America. Paradoxically, the genus *Equus* went extinct in Central and North America after the end of the ice age, around 8,000 years ago, but survived in the Old World. Columbus, Cortez, and the conquistadores brought horses back to their original homeland aboard their ships (Figure 8.3d). In the Americas, horses, together with cannons and guns, helped subdue the original inhabitants, even as the American Indians were becoming skilled horse riders. Until the end of the seventeenth century, *Equus*—now as *E. caballus*—again spread over North America and down into South America. The horse *E. caballus*, which proved to be an ideal transport over rough terrain, became an essential part of Indian cultures (see Figure 2.11).

IMPORTANT SITES OF FOSSIL HORSES are found in the American West, particularly in the area of the Rocky Mountains (Figure 8.4). Unlike in Europe, there are large inner-continental basins, such as the Clarks Fork, Bighorn, Wind River, and Uinta basins (Figure 8.5). Between the Wind River Basin in the north and the Uinta Basin in the south there are three basins further intercalated from west to east: the Wasatch, the Bridger, and the Washakie basins. The bottoms of these inner-continental basins subsided continuously during the Early Eocene. Several hundred to 1,000 meters of sediment, mostly fluviatile, accumulated and then was worn away in these basins, which became known as the Badlands (Figure 8.6). Sometimes the sedimentation process stood still because the rivers meandered. In those areas, fossil soils (paleosols) developed and fossils accumulated in them. Among these fossils were thousands upon thousands of horse remains, particularly jaws, isolated teeth, and bones (Figure 8.7). Complete skulls and even whole skeletons were also deposited (Figures 8.8, 8.10, and 8.11).

The abundance of the material coming from a very dense sequence of fossiliferous layers permits the detailed reconstruction of the evolution of dentitions of certain species. Philip Gingerich from the University of

Figure 8.5. Map of the inner-continental Eocene basins in the Rocky Mountains.

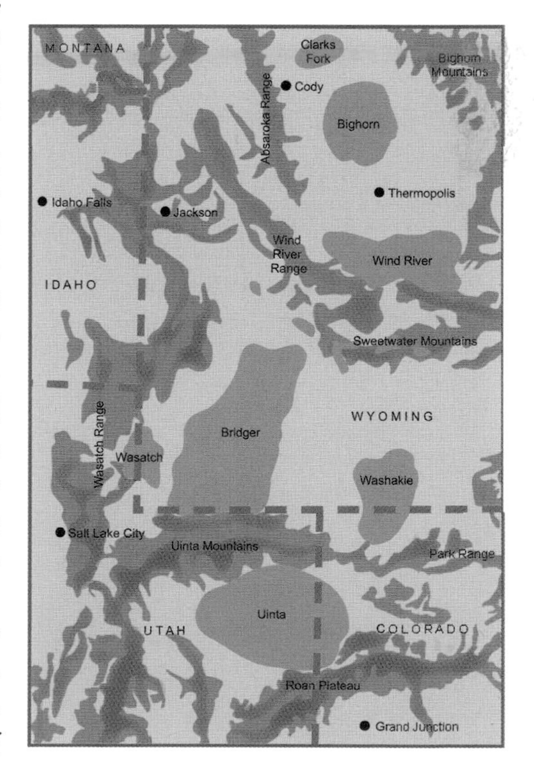

Michigan was even able to trace the continuous development of body weight throughout time. These investigations have shown that there are four successive species of *Hyracotherium*. Amazingly, body size develops in a zigzag pattern (Figure 8.9). One can only speculate as to the reasons for that. It is possible that we are not dealing with subsequent members of a single evolutionary lineage. It is also possible that climatic oscillations—leading to changes in the mean annual temperature and in nutritional offerings—played a role. The smallest and oldest species, *Hyracotherium (Sifrhippus) sandrae*, is represented by a skeleton that was reconstructed from the local accumulation of bones that evidently belonged to a single individual. This species was the size of a domestic cat (Figure 8.10).

Other localities with horses from the Early Eocene occur in Colorado. These are sites in the San Juan Basin and the Castillo Pocket in the Huerfano Basin. There are no complete skeletons, but there are many isolated teeth, fragments of jaws and skulls, and isolated bones. Their deposition seems to have resulted from catastrophic events.

Figure 8.6. Mammal paleontologists survey the Badlands of the Early Eocene Elk Creek Facies (Wasatchian) in the Bighorn Basin, Wyoming (August 1980).

Figure 8.7. Jaw and bone fragments of *Hyracotherium*. The finds, which came from one layer of the Willwood Formation (Wasatchian), demonstrate the abundance of Early Eocene dawn horse finds in the area.

Each one offers a temporal cross section through a fossil population. The material from Castillo Pocket stems from at least 32 individuals, representing two different species, primarily *Hyracotherium tapirinum* and the smaller *Hyracotherium vasacciense*. David Kitts (1956) systematically studied the finds, and Philip Gingerich later performed paleobiological analyses. Based on the different sizes of the canines, Gingerich was able to distinguish between females and males.

Situated between the Wind River Basin in the north and the Uinta Basin in the south, the Green River Formation is within the Bridger and the Washakie basins. This formation dates from the upper Early until the lower Middle Eocene. It is around 47 to 52 million years old. The highest parts of it correspond in age with the fossil site of Grube Messel. Contrary to the other basins, the Green River Formation is composed of sediment from three former lakes—Fossil Lake, Lake Gosiute, and Lake Uinta. Fossil Lake, the deepest of them, has provided the most and the best-preserved mammals, including

Figure 8.8. A skull of *Hyracotherium* discovered on an expedition to the Bighorn Basin in 1985. (A U.S. quarter is used for size comparison.)

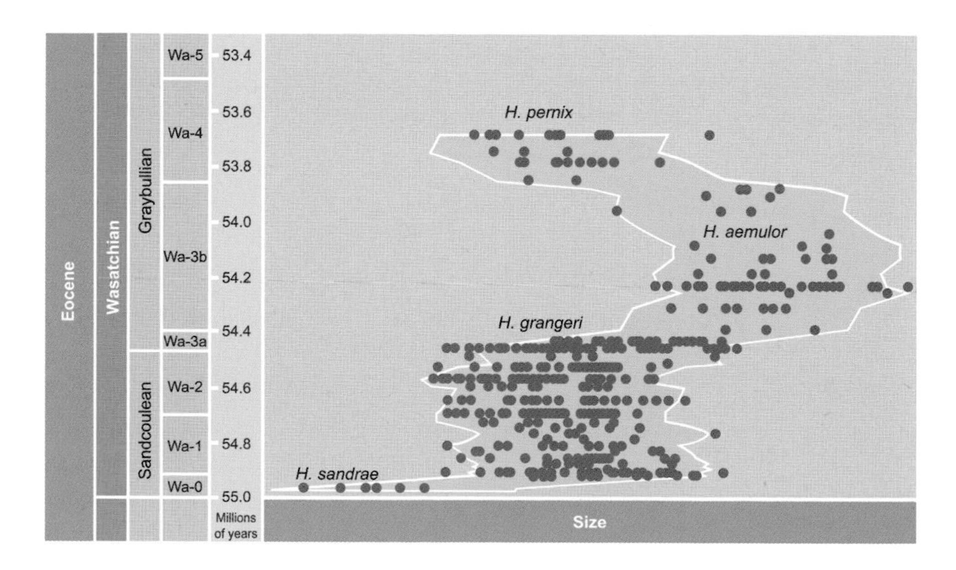

Figure 8.9. Sequence of species and variation of body size in northern Bighorn and Clarks Fork basins *Hyracotherium* during 1.4 million years of evolution. The estimated size (body weight) was based on the size of the first lower molar.

the only articulated skeleton of a dawn horse from the area, a *Hyracotherium* that is on private property (Figure 8.11).

The White River Formation of the Early Oligocene in Wyoming is considerably younger than the deposits of the Green River Formation. It is around 32 to 34 million years old. Its fine-grained, bright gray volcanic ashes have produced numerous excellently preserved mammal skeletons, including several specimens of *Mesohippus bairdi* (Figure 8.12). The cheek teeth of this genus are still so low crowned that the eye sockets are not displaced toward the occiput—as is the case later. Also, the eye sockets are not separated from the temporal fossa by a postorbital bar. The occlusal pattern of the cheek teeth is similar to browsers. The ulna and radius in the forelimbs and the tibia and fibula in the hind limbs are not yet fused. The forefeet and hind feet still consist of three toes, which are, however, already very elongated.

Even younger are fossils of Thomson Quarry from the Early Miocene of Nebraska. This site became renowned during the American Museum of Natural History's 1922 to 1923 excavations. During the

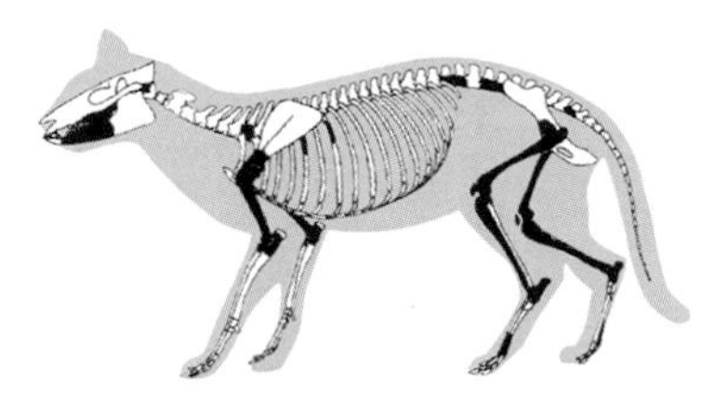

Figure 8.10. Reconstruction of the skeleton of the earliest North American dawn horse *Hyracotherium (Sifrhippus) sandrae* from the Wasatchian 0 level in the Clarks Fork Basin at Polecat Bench, Wyoming (black parts = fossil finds). The skeleton is superimposed over the silhouette of a domestic cat for size comparison.

Figure 8.11. Skeleton of Early to Middle Eocene (Bridgerian C) *Hyracotherium* from Fossil Butte, Wyoming. The beautifully preserved specimen is the only fully articulated Eocene dawn horse from North America. It was preserved in the deposits of a gigantic freshwater lake, which explains the fish fossils seen around the horse skeleton.

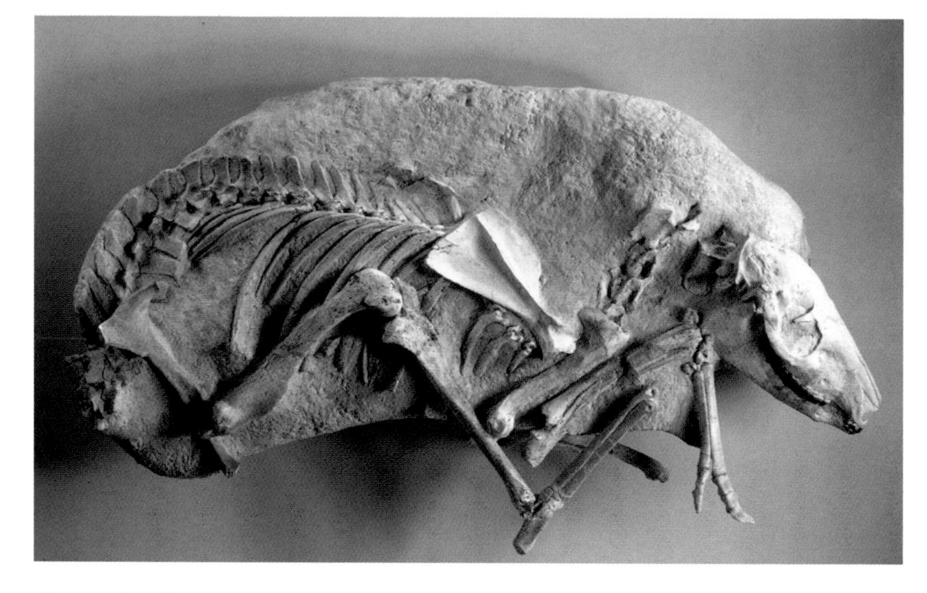

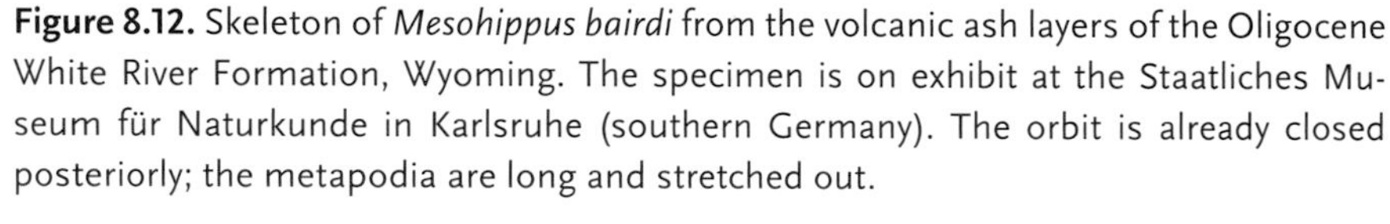
Figure 8.12. Skeleton of *Mesohippus bairdi* from the volcanic ash layers of the Oligocene White River Formation, Wyoming. The specimen is on exhibit at the Staatliches Museum für Naturkunde in Karlsruhe (southern Germany). The orbit is already closed posteriorly; the metapodia are long and stretched out.

Figure 8.13. Skull of *Merychippus primus* seen from below. The teeth already show a clear development toward grazing (relatively high crowned, with tooth cement on the outsides and interspaces of the occlusal surface). On exhibit at the American Museum of Natural History in New York City.

thirties and forties, the Frick Laboratory was also working at the site. Its finds were later turned over to the American Museum of Natural History in New York City. Besides numerous jaws and isolated teeth (Figure 8.13), the collection today contains the fossil remains of no fewer than 150 individuals, including 125 horse skulls. All of these belong to the same species, *Merychippus primus*. They represent unique material for paleobiological studies. Interestingly, these remains involve a critical moment in the evolution of horses, namely the transition from browsing to grazing.

An area with special significance is the John Day Fossil Beds in Oregon, spanning the time period from the Early Eocene to near the end of the Miocene, encompassing much of the Tertiary Period. This makes it possible to isolate a considerable part of the horse evolution in a single area. Its fossil record begins with *Epihippus* and *Orohippus* in the Clarno Formation dating from 37 to 54 million years of age. Following these are *Mesohippus, Miohippus,* and *Kalobatippus* from the John Day Formation *sensu stricto,* from 18 to 39 million years old. *Archaeohippus, Hypohippus, Merychippus,* and *Parahippus,* around 12 to 15 million years old, appear in the Mascall Formation, followed by the Rattlesnake Formation with *Neohipparion* and *Pliohippus* from 6 to 8 million years ago.

Still younger is one of the best sites with fossil horses in all of

Figure 8.14. Excavation quarry in a Rancho La Brea tar pit in Hancock Park, Los Angeles.

North America, the Hagerman Quarry. It was discovered in 1928 near Hagerman, a small village in Idaho. J. W. Gidley and C. Lewis Gazin led the excavation team from the U.S. National Museum (the Smithsonian). The dig yielded 8 to 10 skeletons of dawn horses and more than 130 skulls, many of them with mandibles, of a single species, *Pliohippus shoshonensis* (also known as *Equus simplicidens* and *Equus shoshonensis*). This horse is a descendant of *Dinohippus*. The quality of preservation is outstanding. Even such delicate structures as hyoid bones are present. The site was originally in the vicinity of a small pond, possibly a watering place. Its age is around 3.5 million years. The size of the horses was already similar to modern-day Arab horses, measuring 425 kilograms on average.

Among the sites for fossil horses in North America, one of the most famous is the La Brea tar pits in the heart of downtown Los Angeles (Figure 8.14). The site includes remains of two horse species, the larger *Equus occidentalis* and the somewhat smaller and more slenderly built *Equus lambei*, which is also known as *Equus conversidens* (see Figure 10.8). These were the last genuine horses of

Figure 8.15. Entrance of the George C. Page Museum in Hancock Park, Los Angeles. The frieze above the entrance depicts life in this area at the end of the ice age.

North America. They went extinct some 8,000 years ago. In addition to horses, many other fossils are preserved at La Brea, including plants, invertebrates, and vertebrates. Mammoths (*Mammuthus columbi*), American mastodons (*Mammut americanum*), saber-toothed cats (*Smilodon fatalis*), and dire wolves (*Canis dirus*) are among the many mammal fossils at the tar pits. In total no fewer than 565 species are documented. The diverse fossils allow a comprehensive reconstruction of the biotope that represents the youngest North American mammal age, the Rancholabrean. Radiocarbon dating has proved that the series of small tar pits spans a time interval from 40,000 to 4,000 years BP. The presentation of fossil finds in the George C. Page Museum and continuing excavations and other investigations at Hancock Park are exemplary (Figure 8.15).

Asia

Asia has not produced a candidate for the origin of the horse's family (Equidae), apart from *Radinskya yupingae,* whose systematic and phylogenetic relationships are unclear. In Asia, as in Europe, there is no continuous record of the horses' evolution. The oldest finds are said to come from China and consist of two fragments of lower jaws with teeth that were discovered together with a series of other mammal species in the upper Lingcha horizon of the Henyang Basin in southwest China. These fossils are said to correspond in age with the oldest Eocene mammal fauna of North America (from Wasatchian). However, inconsistencies are apparent in the determination of the fossils. For example, the genus *Propachynolophus* is biochronologically younger than *Hyracotherium*, from which it should have developed. *Propachynolophus* occurs in Europe no earlier than the upper Early Eocene. It is therefore interesting that *Propachynolophus hengyangensis* from China was recently considered together with some other alleged equid species, such as *Propalaeotherium sinense* and *Gobihippus gabuniai*, as a special group to be more likely related to chalicotheres (see Chapter 9) than to equids. If this holds true, there would be no dawn horses known from the Early Eocene of Asia, and thus the problem of the origin of the horses' family would be simplified (see Chapter 11).

In that case, some upper and lower jaws with teeth—described as *Qianohippus magicus* from the Late Eocene or Early Oligocene of the Shinao Basin in the southwest of China—would be the first record of fossil horses from Asia. Obviously, however, this species did not initiate the development of an Asian branch of horses during the Oligocene. It was not earlier than the Early Miocene that the next equid, *Anchitherium aurelianense*, appeared in Asia. Obviously, it migrated from North America by way of the Bering land bridge. Soon it expanded all over the Asian continent and into Europe. In China, a species of its own developed—*Anchitherium zitteli* (Figure 8.16). During the Late Miocene, *Anchitherium* was replaced by hipparions, initially by *Hippotherium* but then a few million years later by *Hipparion* itself. In the beginning, *Hipparion* took the same northern route as *Anchitherium* and *Hippotherium*; then it turned southward and appeared around 9 million years ago south of the Himalayas in the layers of the Siwaliks. Interestingly, *Sinohippus*, a browsing descendant of *Anchitherium*, appeared about the same time with the grazer *Hipparion*. It remained restricted to East Asia where it reached a gigantic size.

Figure 8.16. Radiation and evolutionary lineages of *Anchitherium* during the Miocene of Eurasia. The names on the graph represent key locations with *Anchitherium*.

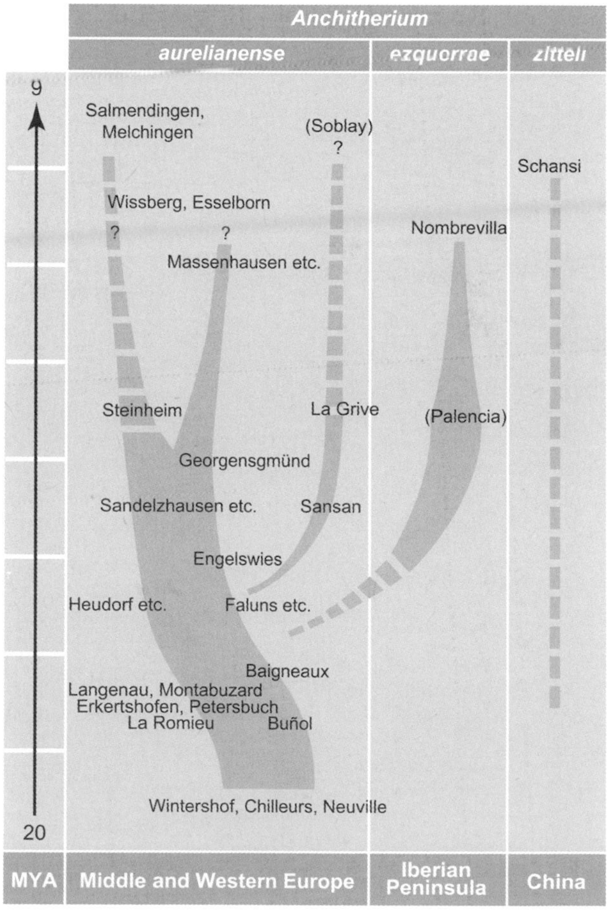

The next immigrant to Asia was the genus *Plesippus*. Derived from the North American *Plesippus shoshonensis*, it appeared shortly before the ice age, around 2.5 million years ago. Among its earliest representatives are the skulls of the very big *Plesippus eisenmanni*, which was discovered in 2-million-year-old layers of the Lingcha Basin in northwest China. The first representative of the genus *Equus* in the Old World was *E. granatensis*. It appeared in Asia around 1.5 million years ago. This species was completely isolated. There are no phylogenetic connections known either to *E. coliemensis* or *E. nalaikhaensis*—which is about half a million years younger—or to the oldest representatives of the half-asses (see Figure 10.8).

Asia is very important in modern times with respect to the history of horses. The last wild-living representatives of modern horses, *Equus przewalskii*, were observed in the Mongolian steppes during the 1960s. Contrary to former assumptions, this species is not directly related to our domestic horses (*Equus caballus*) (see Chapter 10). In addition to *E. przewalskii*, Asia is also where half-asses of the species *Equus hemionus*—such as onager, djiggetai, kulan, khur, and kiang—are still living wild in the deserts and semideserts of the Near East up to the highlands of Tibet.

Europe

Europe seems to have been the northern continent first occupied by direct ancestors of our modern equids. North America's location simply prevented its being reached by immigrating horses from Africa or India. The few Asiatic Eocene horses were rather isolated, and they don't display any transitional forms linking European to North American hyracotheres or vice versa (Figure 8.3a). Since Richard Owen's initial description of the genus *Hyracotherium* in 1840, there have been more finds of this genus from Europe, although most of them consist only of isolated teeth. Those remains were discovered mainly in England, the Paris Basin, and southern France. Unlike North American skeletal remains, these specimens are lacking in limb anatomy that could offer information about their body construction.

Several genera developed from *Hyracotherium* in Europe during the Early Eocene, for example, *Propachynolophus*, *Pachynolophus*, and *Lophiotherium*. During the Middle Eocene, *Propalaeotherium* appeared as a descendant of *Propachynolophus*, while *Eurohippus* developed out of *Pachynolophus*. The genus *Anchilophus* was limited to southern and western Europe. Its origin is unknown. It appeared no earlier than the Middle Eocene and as late as the Late Eocene (see Figure 11.1). There is still debate about whether *Anchilophus* is an equid at all. Numerous skeletons of *Propalaeotherium* and *Eurohip-*

pus have been discovered in the Messel Pit near Darmstadt; a few were also discovered at Eckfeld and in the Geiseltal (see Chapter 5). Based on these discoveries, we know a lot about the body construction of these two genera. A comparison with the skeletal finds from North America demonstrates that *Propalaeotherium* and *Eurohippus* still looked rather similar to *Hyracotherium*. The main difference was in the further development of the dentition. The forelimbs still bore four hooves and the hind ones three. The back was also highly arched, similar to modern duikers and muntjacs. Reconstructions give us an idea of what these early dawn horses looked like, how they behaved, and how they moved in their natural environment, a luxurious tropical rain forest (see Figures 5.27 and 5.28).

Similarly built to *Propalaeotherium* and *Eurohippus* was the genus *Hallensia*. Its dentition looks much more primitive; its bumpy occlusal pattern resembles that of phenacodonts, North American dawn ungulates (Condylarthra). Evidently, *Hallensia* represents an early side branch in the evolution of dawn horses that was obviously derived from the same source as *Hyracotherium*, although independently (see Figure 11.1). It is only known from Europe. It corroborates the hypothesis that the equids immigrated into the northern continents by way of Europe. In Europe *Hallensia* appeared contemporaneously with *Hyracotherium* at the beginning of the Eocene. It went extinct during the Middle Eocene. Four genera of hyracotheres developed in Europe by the Middle Eocene. These were *Propalaeotherium, Eurohippus, Lophiotherium,* and *Anchilophus. Propalaeotherium* had already disappeared at the beginning of the Late Eocene, while *Eurohippus* and *Lophiotherium* survived well into the middle of that epoch. Only *Anchilophus* survived until the end of the Eocene. During the Late Eocene, the equids were being replaced by the palaeotheres, which appeared during the Middle Eocene in Europe. These are still confused by many with true equids. Where these pseudo horses came from is still unknown (see Chapter 9). Recent discoveries like *Bepitherium* from the Early Eocene of Spain point, however, to an African origin by way of the Iberian Peninsula.

During the time interval from the upper Late Eocene until the end of the Early Miocene, or between 19 to 35 million years ago, there were no horses in Europe at all. It was only with *Anchitherium* that the horses reappeared in Europe. *Anchitherium* came from North America by way of Asia. It was a middle-size browser with three toes on the front and hind legs and low-crowned teeth (Figures 8.3b, 8.16, and 8.17). The anchitheres existed in the form of *A. aurelianense* in Europe up until the Late Miocene without changing very much, except for the development of several subspecies. It was only on the Iberian Peninsula that a separate species, *Anchitherium ezquerrae,* developed.

Interestingly, the last survivors of *Anchitherium* met the next wave of horse invasions from North America around 11 million years ago (Figure 8.3c) represented by *Hippotherium primigenium.* This three-toed horse already displayed relatively high-crowned cheek teeth (see Figures 6.2 and 6.13). *Hippotherium* arrived in the Old World via the Bering land bridge as its close relative *Hipparion* would do 2 million years later. Another relative, *Allohippus,* would do the same shortly before the beginning of the ice age, about 2.5 million years ago, and would immigrate into Europe at the time *Hipparion* went extinct.

The oldest representative of the genus *Equus* was the large *E. suessenbornensis;* it appeared in Europe around 1 million years ago (see Figure 10.8). Its origin is still unknown. It could possibly be related to the more or less contemporaneous Asiatic species *E. coliemensis* and *E. nalaikhaensis.* In spite of size similarity between the species, there was no transition to the Middle Pleistocene *E. mosbachensis,* which is considered to be the oldest ancestor of modern horses (see Figure 10.9).

Except for Messel, the Geiseltal, and Eckfeld (see Chapter 5), other important fossil horse sites in Europe correspond to later stages of the evolutionary development of the horses. Similar to the Paleogene of North America, large inner-continental basins developed in Europe during the Miocene in connection with Neogene orogenesis, in this case the development of the Alps. Like the Bavarian and Swiss Molasse basins, they were filled with alternating marine and terrestrial sediments. The only horse species of those times that occurred in these basins north of the alpine mountain range was *Anchitherium aurelianense.* Numerous dental and bony fragments of this species have been found at many sites in the upper Bavarian Molasse, which represents the debris of the eroded uplifting Alps (Figure 8.17).

A truly exceptional site containing the Late Miocene *Hippotherium primigenium* is the Höwenegg locality, situated in the Hegau region, northwest of Lake Constance (Figure 8.18). Since the early 1950s, excavations have been undertaken in the limnic sediments at the foot of a former volcano, initially by the Geologisch-Paläontologisches Institut of the University of Freiburg im Breisgau, then by the Natural History Museums of Karlsruhe, Darmstadt, and Stuttgart. No fewer than 15 skeletons of *Hippotherium primigenium* have

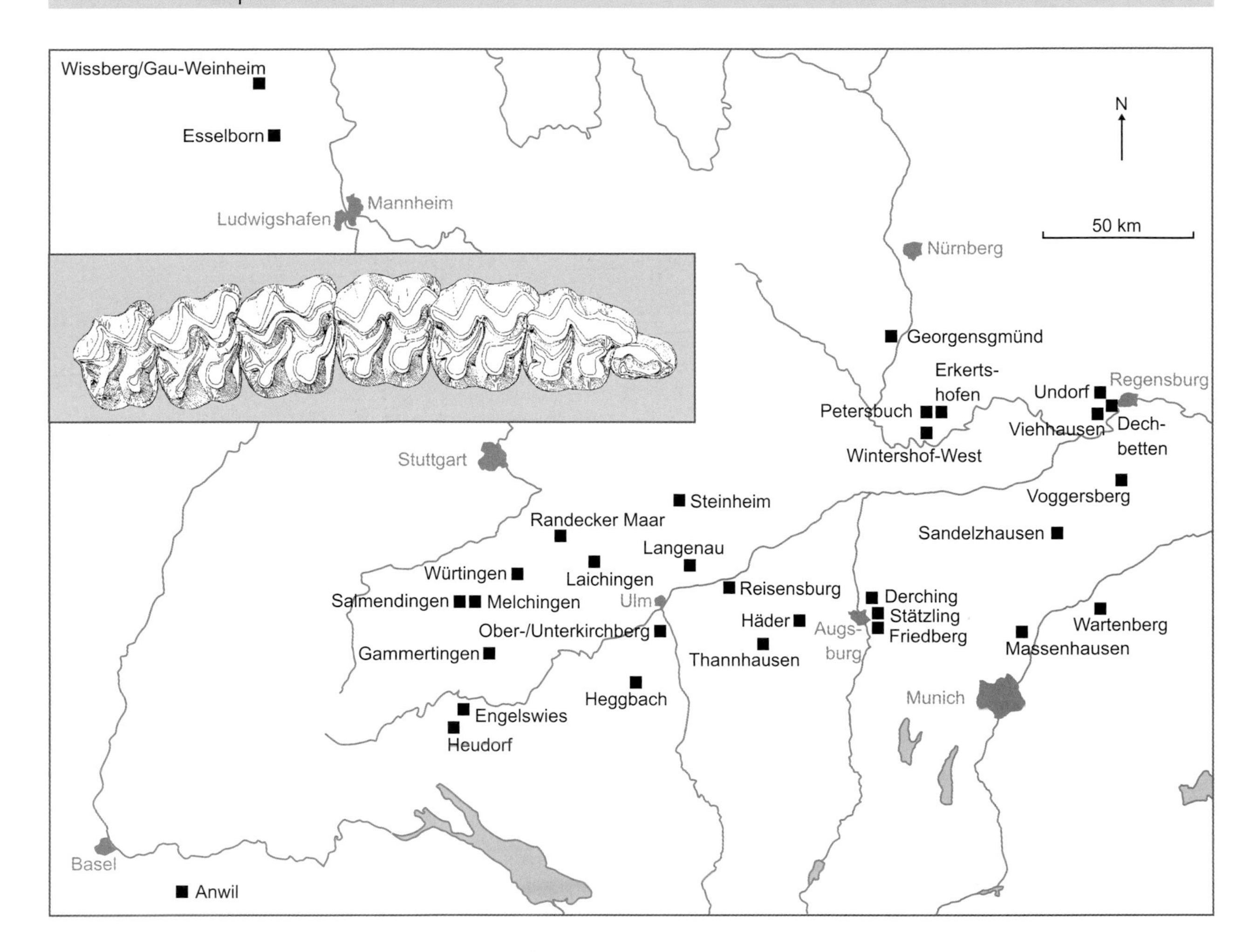

Figure 8.17. Sites where *Anchitherium aurelianense* fossils were found in southern Germany and Switzerland. *Inset*: a maxillary dentition of this species from Sandelzhausen, Bavaria.

thus far been discovered, including a fetus inside a mare. These finds, together with a mammal fauna and some plants of the Late Miocene, may have been preserved in the sediments of a former maar lake, like Messel and Eckfeld. This is still a matter of ongoing debate and excavations at this site continue. The skeletons of *Hippotherium primigenium* are the only ones known from this genus. They were the subject of a comprehensive monograph prepared by a joint American-German team consisting of Ray Bernor, Heinz Tobien, Lee-Ann Hayek, and Walter Mittmann.

The first finds of *Hippotherium primigenium* (considered for a long time to be *Hipparion*) were discovered at the beginning of the nineteenth century near the small village of Eppelsheim in Rhein-

Figure 8.18. Site of the Staatliche Museum für Naturkunde of Karlsruhe and Stuttgart dig in the lake deposits at the foot of the extinct Höwenegg volcano in the Hegau region north of Lake Constance (southern Germany).

hessen (Figure 8.19). The isolated jaws, teeth, and bones were described and named in 1829 by Hermann von Meyer, the vertebrate paleontologist at the Senckenberg Museum in Frankfurt at the time. These finds, however, do not comprise full skeletons. Important fossils of the somewhat younger *Hipparion*, mainly numerous skulls, were found during the nineteenth century in Greece, at Pikermi near Athens and on the island of Samos. Several well-preserved skulls and other material of *Hippotherium*, as well as *Hipparion*, have been found in Spain.

There are also important sites of *Allohippus* known from Europe. Among them is the Late Pliocene locality of Senèze, situated in the central plateau of France. A complete skeleton of *A. stenonis* (see Figure 10.7) has been found there. Characteristics of this species have been considered indicative of direct connections with the zebras, which are otherwise only known from sub-Saharan Africa. Currently, however, these features are more likely considered as parallel or convergent developments.

A complete skeleton of the large Middle Pleistocene *Equus mos-*

Figure 8.19. Depiction of the Eppelsheim site, where a skull of the proboscidean *Deinotherium giganteum* was salvaged in 1835. The first relics of the three-toed dawn horse *Hippotherium primigenium* were also found here.

bachensis was reconstructed from numerous isolated bones collected from the Mosbach Sands, sediments of the Rhine of that time (see Figure 10.9). These sands and gravels in the region between the cities of Mainz and Wiesbaden are still being investigated. Horses of the late ice age appear in cave paintings in southern France, northern Spain, and south of the Ural Mountains in eastern Russia. Here it is possible to look at ancient horses through the eyes of our ancestors (see Figure 10.10).

Africa

With the exception of one bat from Mahenge in Tanzania, no site south of the Sahara has provided Paleogene mammals. Consequently, for the period from 23 to 65 million years ago, we know nothing of mammal faunas that lived in this vast area. Specifically, it is not known whether horses and their ancestors once thrived in Africa, although the earliest origin of equids could indeed be that continent. The oldest horse fossils known from Africa are for the

Figure 8.20. Differentiation and expansion of the zebras in East and South Africa.

genus *Hippotherium*, which clearly came from Asia and appeared no earlier than 10 million years ago, during the Late Miocene. However, it got no farther than North Africa (site of Bou Hanifia in northern Algeria). Possibly, the way to the south was blocked during this time by a desert or at least a dry area of the Sahara. A few million years later *Hipparion*, which was a grazer and therefore possibly better equipped for such a passage, succeeded in making the journey south. Representatives of this genus pushed forward to East Africa where descendants in the form of *Stylohipparion* survived up into the Middle Pleistocene (Figure 8.3c).

Crossing the deserts or semideserts of the Sahara area was not impossible for grazing horses. *Allohippus* arrived south of that barrier in East Africa during the Pliocene (see Figure 10.8). The first representative of the genus *Equus* in Africa, *E. tabeti*, is only known from an isolated occurrence in the early Lower Pleistocene of North Africa. During the middle of the ice age, *Equus mauretanicus* appeared in North America and *E. capensis* in South Africa, leading to the later development of the steppe zebras (*E. burchelli*, Figure 8.20; see also Figure 10.8). The origins of the Grevy's zebra (*E. grevyi*) and mountain zebras (*E. zebra*), however, are still not clear. Another branch of *Equus*—*E. melkiensis*, known from the Middle to the Late Pleistocene of North Africa—developed into the wild asses (*Equus asinus*; see Figure 10.8). Zebras and asses, of course, still live in the African wilderness.

South America

During most of the Cenozoic—the modern age of earth's history—the South American subcontinent was completely isolated from the rest of the world. Ungulates did not originally occur there, but specialized hoofed mammals evolved (see Chapter 9). About 2 to 3 million years ago a significant faunal interchange (the Great American Biotic Interchange) occurred between North and South America by way of the isthmus of Panama. During this exchange, the first true horses arrived in South America. The earliest South American horse fossils are of *Hippidion*, dating to 2.5 million years ago. It is still unknown if this genus developed in South America or if it was already present in the southern parts of North America shortly before the first migrations took place. Compared to *Equus*, *Hippidion* is characterized by its smaller size and by its very deep nasal incision, which is connected to reduced nasal bones that are narrow, long, and stretched-out splints of bone (see Figure 10.6). Strongly shortened fetlock bones are viewed as an adaptation to the mountainous biotope of the Andes. Based on genetic analyses, *Hippidion* developed from *Equus* no earlier than the end of the Pliocene (see Figure 10.8). From a classic paleontological point of view, however, *Hippidion* is more likely a descendant of *Pliohippus*. Of course, the migrations between North and South America did not take place in only one direction. So *Hippidion* is also regarded as a re-immigrant appearing in the Pleistocene of Southern California where it allegedly previously developed from *Dinohippus*. In any case, the evidence is not conclusive. As a parallel example, there are quite a number of fossil ground sloths known from Pleistocene deposits in the southern United States that confirm immigrations of South American mammals into North America.

The genus *Equus* arrived in South America around 2 million

years ago, a short time after *Hippidion*. There, *Equus* gave rise to several lineages, including *E. andium* (see Figure 10.8). Some place these species in a genus or subgenus of their own under the name *Amerhippus*. *Equus andium* is characterized by conspicuously short metapodia. Their development—like that of the shortened fetlocks of *Hippidion*—is considered to be connected with living in the high mountains of the Andes. Unknown until recently was the amazing fact that both *Hippidion* and *Equus andium* survived well into post-glacial times in Southern Patagonia.

Australia

Australia was the last continent reached and inhabited by horses. Horses arrived in Australia and New Zealand with the first ship of European immigrants in 1788. Since then, horses have spread all over the continent at amazing rates. Today there are more than 200,000 horses on the continent.

CHAPTER NINE

Pseudo Horses and Relatives of Horses

HORSES BELONG to the mammalian order of odd-toed ungulates or Perissodactyla (from Greek *perissos* = odd, *dactylos* = finger). This order was once as dominant on the planet as the even-toed ungulates (Artiodactyla, from Greek *ártios* = even) of today. During the Eocene and Oligocene, the perissodactyls represented the majority of ungulates, in total number of individuals, but not in genera or families. About 15 families of perissodactyls existed, compared to about 30 families of artiodactyls. The number of genera was around 80 percent of that of artiodactyls. Today the rhinos, the tapirs, and, of course, the horses are the last surviving perissodactyls. As a group, perissodactyls are becoming more and more endangered, with the exception of some species that live in remote areas. On the other hand, numerous artiodactyls can be seen on the savannahs and steppes of East Africa, including buffalos, gnus, giraffes, hippos, and countless species of antelopes. In contrast, aside from the rather successful zebras, Africa's perissodactyls are represented only by a few rhinos. What might be the reason for the modern paucity of perissodactyls?

Usually, the evolutionary victory of the artiodactyls over the perissodactyls is explained by touting the superiority of rumination (digestion in the foregut) over cecum fermentation. However, the American paleobiologist Christine Janis has shown that this is not true. Cecum fermentation and rumination (which represent different digestion strategies) are both strategies used to ferment cellulose-rich (and therefore indigestible) food with the help of bacterial colonies. The difference is that in cecum fermenters, the digestion of cellulose takes place in the hindgut and consequently distal to the small intestine, which is where most of the products of digestion are absorbed. Conversely, in ruminants digestion takes place in the foregut and therefore proximal to the small intestine (Figure 9.1). In cecum fermentation, many of the products cannot be assimilated into the body. Therefore, rumination is ostensibly more effective. But the digestive apparatus in ruminants makes up 40 percent of their body weight compared with only 15 percent in cecum-fermenting horses. This in turn can be decisive for the survivability of cursorial (running) animals such as horses. Moreover, cecum fermenters are able to process the dry parts of cellulose-rich stems, while ruminants depend more heavily on lush, cellulose-poor food (Figure 9.2). This is why horses, zebras, half-asses, and asses were able to expand their biotopes into the dry steppes and semideserts where ruminants cannot successfully compete. An exception among the artiodactyls are camels, which possess a different kind of digestive system. On savannahs, where plenty of cellulose-poor food is available, the digestive system of ruminants works more effectively than that of cecum fermenters. This is why they appear superior in those areas. An interesting case is the zebra, which frequently occurs in savannah

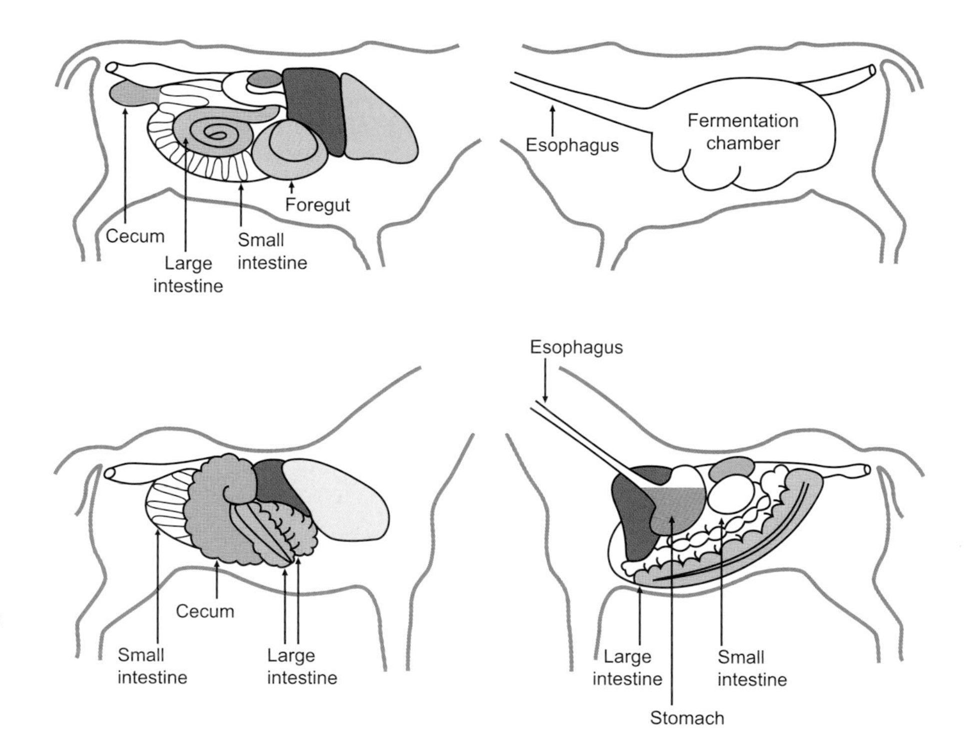

Fig 9.1. Comparison of the digestive systems of ruminators and cecum fermenters as exemplified by a cow (*top*) and a horse (*bottom*). Reddish brown = liver; light gray = lung; dark gray = kidney.

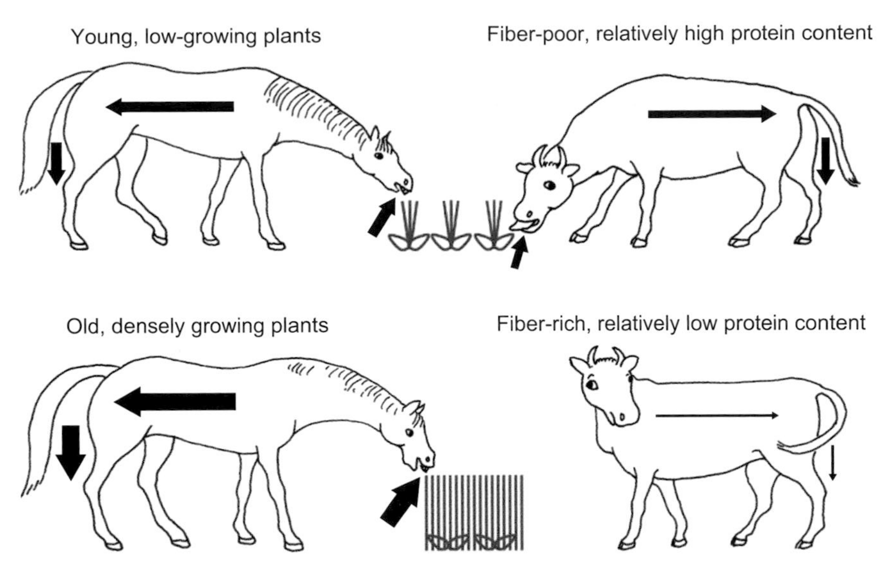

Figure 9.2. Comparison of the feeding strategies of cecum fermenters, as exemplified by a horse (*left*), and ruminants, as exemplified by a cow (*right*). *Top*: In the case of young, low-growing plants, ruminants feed primarily on leaves, which are fiber poor and protein rich. Cecum fermenters are content with the stem, which is fiber rich and protein poor. Horses have to take in much larger amounts of nutrition than cows. The efficiency of cecum fermentation is only 70 percent of that of ruminants. *Bottom*: In the case of old, densely growing plants with high fiber and low protein content, the situation favors the cecum fermenters because they retain the same digestion efficiency whereas the digestion efficiency of ruminants decreases to an extent that it is no longer satisfactory for providing nutrition. Horses can thrive in dry areas such as steppes and semideserts; ruminants cannot.

biotopes. In contrast to gnus, which feed on the same kinds of plants, zebras are content with stalks and stems whereas the gnus need the leaves. This food preference allows them to live side by side without competing with each other.

Today's tapirs and rhinos are like living fossils (Figure 9.3). With their four-hoofed forelimbs and three-hoofed hind limbs, the tapirs represent a level of evolutionary development that the horses left behind with *Mesohippus* toward the end of the Eocene, about 37 million years ago. At about the same time, rhinos reduced the number of their hooves on the forelimb to three. However, neither tapirs nor rhinos made it to solid hoofedness. The reason such a seemingly primitive locomotory system was retained could have been because rhinos developed massive bodies, which needed extra support. Three-digital extremities would have been better suited than the slender and comparatively sensitive legs of modern horses. In the case of tapirs, locomotion on soft forest bottoms could have played a decisive role, although they also developed rather massive bodies.

Most but not all tapirs today live in tropical rain forests. The mountain tapir (*Tapirus pinchaque*) lives up to 4,500 meters in the Andes, where it is able to exist at rather low temperatures. The other American tapir species—*Tapirus terrestris* and *Tapirus bairdi*—can also survive in mountainous areas. The Asian or Malayan tapir—*Tapirus indicus*—typically lives in the tropical rain forests of Burma

Figure 9.3. Representatives of living and extinct perissodactyls (*top to bottom*): *Brontotherium* represents the extinct Brontotheriidae; *Moropus*, the extinct Chalicotheriidae; *Equus* the present-day Equidae; *Palaeotherium*, the extinct Palaeotheriidae; *Hyracodon*, the extinct Hyracodontidae; *Tapirus*, present-day Tapiridae; *Metamynodon*, the extinct Amynodontidae; *Ceratotherium*, the present-day Rhinocerotidae.

(Union of Myanmar) and Thailand up to Malaysia and Sumatra. A solitary being, it walks during twilight hours on beaten tracks through the undergrowth alongside rivers. The pattern of its hide—black on its head, the front of its body, and its forelimbs, and white gray on its hindquarters—makes for an excellent camouflage. Tapirs never inhabited steppes and savannahs because they would not have found enough leaves, the major item in their diet. Tapirs never developed high-crowned cheek teeth, and their low-crowned cheek teeth are not suited for hard grasses. Their body construction, dentition, the fact that they inhabit areas far apart from each other—South and Central America on one side and Southeast Asia on the other—all indicate a rather old geological age for this animal group. Like the oldest dawn horses, the first tapirs existed in the earliest Eocene—*Cymbalophus* in Europe and *Cardiolophus* in North America. Both were morphologically rather close to the oldest equid, *Hyracotherium*. They differed from the somewhat later, but still early Eocene dawn tapir *Homogalax*, by having low-crowned, less lophodont, and altogether more primitive dentition. If *Radinskya* from the Late Paleocene of China belongs to this group, then tapiroids would be the oldest perissodactyls that are documented by fossils. In any case, their skeleton—as it is known from *Hyrachyus* at Messel (see Figure 1.4), but also from other species of this genus in North America—shows a marked resemblance to the dawn horses of that time, which in turn would indicate a close phylogenetic relationship with them.

The extinct lophiodonts (from Greek *lophos* = hill, *odóntes* = teeth), which are classified with the tapiroids, were restricted to Europe. The occlusal surface of the cheek teeth of these animals consisted of low, hill-like yokes (Figure 9.4). Their structure resembled that of modern tapirs as well as early relatives of rhinos. The deep nasal incision connected to a trunk, which is typical for later tapirs, is missing. Lophiodonts appeared in Europe for the first time toward the end of the Early Eocene, around 50 million years ago. They may possibly date back to the genus *Homogalax*, indicating a North American origin of this group. During the Middle Eocene, developments took place along many pathways, culminating in *Lophiodon rhinocerodes* of the middle Late Eocene, which had already achieved the size of modern horses. Their body construction, however, in no way resembled that of horses. Unlike horses, lophiodonts were rather thickset, with a massive skull (Figure 9.5). Sites such as the Geiseltal in Germany or Robiac in southern France prove that these animals were abundant in certain, presumably humid, biotopes. Therefore, it is all the more surprising that fossil evidence of this group is ex-

Figure 9.4. Upper jaw of *Lophiodon* from the late Early Eocene of Monthelon (Paris Basin). A complete set of cheek teeth is seen from their occlusal surface; the alveola of a canine is also visible. The original is housed in the Naturhistorisches Museum in Basel (Switzerland).

tremely rare at Messel. The virtual absence of fossils could be a consequence of their large body size; at Messel only small creeks flowed into the lake. These creeks were presumably too weak to transport carcasses of adult-size lophiodonts. This is supported by the fact that at this point the only skeleton of a lophiodont discovered at Messel is not an adult but rather a juvenile, with a length of about 1 meter. Its carcass was obviously small enough to not be filtered out.

Most of the separate, multiple developments of tapiroids occurred during the Eocene of Asia. Lophialetidae and Deperetellidae only existed on that continent. By contrast, the Helaletidae and Isectolophidae families were present in Asia and in North America, with the helaletids also being present in Europe (Figure 9.6). In the course of 50 million years, the tapirs developed in North America through genera such as *Heptodon* and *Helaletes* of the Eocene, *Protapirus* of the Oligocene, and *Miotapirus* of the Miocene, finally resulting in the genus *Tapirus*. This development was characterized by an increasingly deep nasal incision combined with

Figure 9.5. Skeleton of a juvenile *Lophiodon* from Messel. The skull is massive, and the metapodia are unusually short. A thick, dark package of gut content is visible to the right, between the ends of the ribs and the pelvis. The exhibit can be seen at the Senckenberg Museum (Frankfurt am Main).

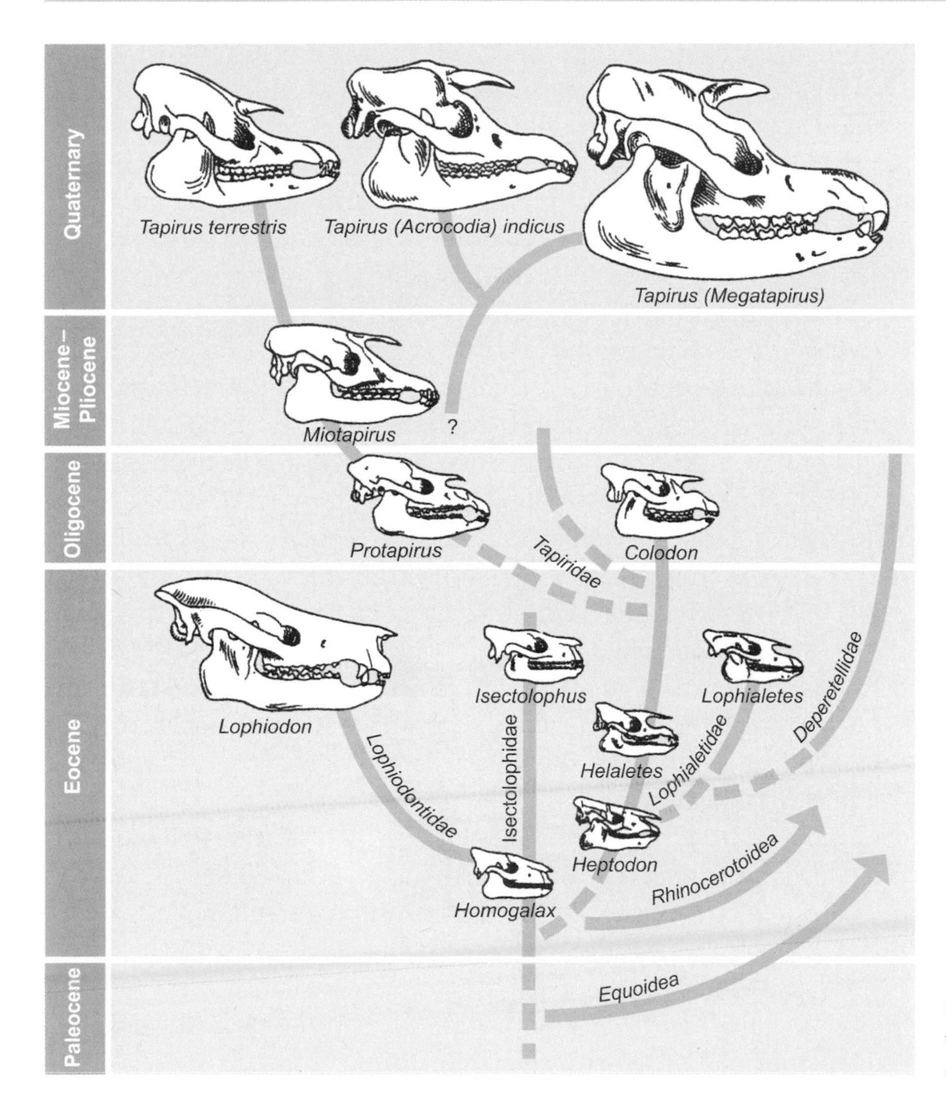

Figure 9.6. Family tree of the tapiroids shown by skull development.

a reduction of the nasal bones. Both of these developments were connected with the formation of a trunk, which is typical of modern tapirs. Only a few changes occurred in the body skeleton, which were obviously due to increases in body size and a certain trend to improve locomotion, even though the limbs became proportionally shorter.

In addition to horses and tapirs, the third group of perissodactyls that still exists today are the rhinos. Four genera with five species are actually known from Africa and Asia. Some of them—such as the Indian rhinoceros (*Rhinoceros unicornis*), the Sumatran rhinoceros from Sumatra and Borneo (*Dicerorhinus sumatrensis*), and the Javan rhinoceros of Southeast Asia and Java (*Rhinoceros sondaicus*)—are inhabitants of tropical rain forests. Others—such as the white rhinoceros (*Ceratotherium simum*) and the black rhinoceros (*Diceros bicornis*) from Africa—ventured into the biotopes of the wide East and South African savannahs. Whereas the white rhino prefers moist underground with junglelike vegetation, the black rhino prefers forested steppes. Different ways of obtaining food led to different postures of the head, more upward-oriented in the browsers, such as the black rhino, and more downward-oriented in the grazers, such as the white rhino. The acutely projecting upper lip of the black rhino serves for plucking off leaves; the broad mouth of the white rhino is better suited for grazing. Both modern African rhinos have two horns on the bridge of the nose, one behind the other, the anterior of which is bigger and more pointed, particularly on the black rhino. The single horn of the Indian and the Javan rhino is blunt, but rather massive in the Indian rhino. The Sumatran rhino is the only Asiatic rhino that bears two horns; the posterior one is often very small. In contrast to the horned artiodactyls, the horns of the rhinos form from the epidermis. They do not contain any bony support, whereas the artiodactyls display at least two bony cones, which bear antlers in deer. The anterior horn of the rhinos can be as sharp as a dagger and weigh several kilograms.

Like the tapirs, the rhinos date far back into the Paleogene in their phylogenetic development. During the Early Eocene, the rhinos diverged from the lineage leading to the genus *Heptodon* (Figures 9.6 and 9.7). The bifurcation possibly occurred with the genus *Hyrachyus*, which is also known from Europe since the late Early Eocene. In any case, it is possible to trace all the multiple developmental lineages of the superfamily Rhinocerotoidea (from Greek *rhinos* = nose, *ceros* = horn) to *Heptodon* and *Hyrachyus*. As with the horses, many far-reaching migrations took place. These were gradual shifts in the population distributions that occurred over many generations. Over the span of geologic history, those shifts achieved continental dimensions.

A geographic shift occurred during the Middle Eocene when some offspring of *Hyrachyus* ventured over the Bering land bridge from Asia to North America. Out of this group developed the genus *Amynodon*, an odd-toed ungulate that was the size of modern tapirs. It differed from the other perissodactyls of that time by having big canines, a movable upper lip, and square-shaped last upper cheek teeth. Two groups developed from Asiatic representatives of this genus. One remained in Asia and culminated with the genus *Cadurcodon*, which later made it into Europe. Being browsers, they devel-

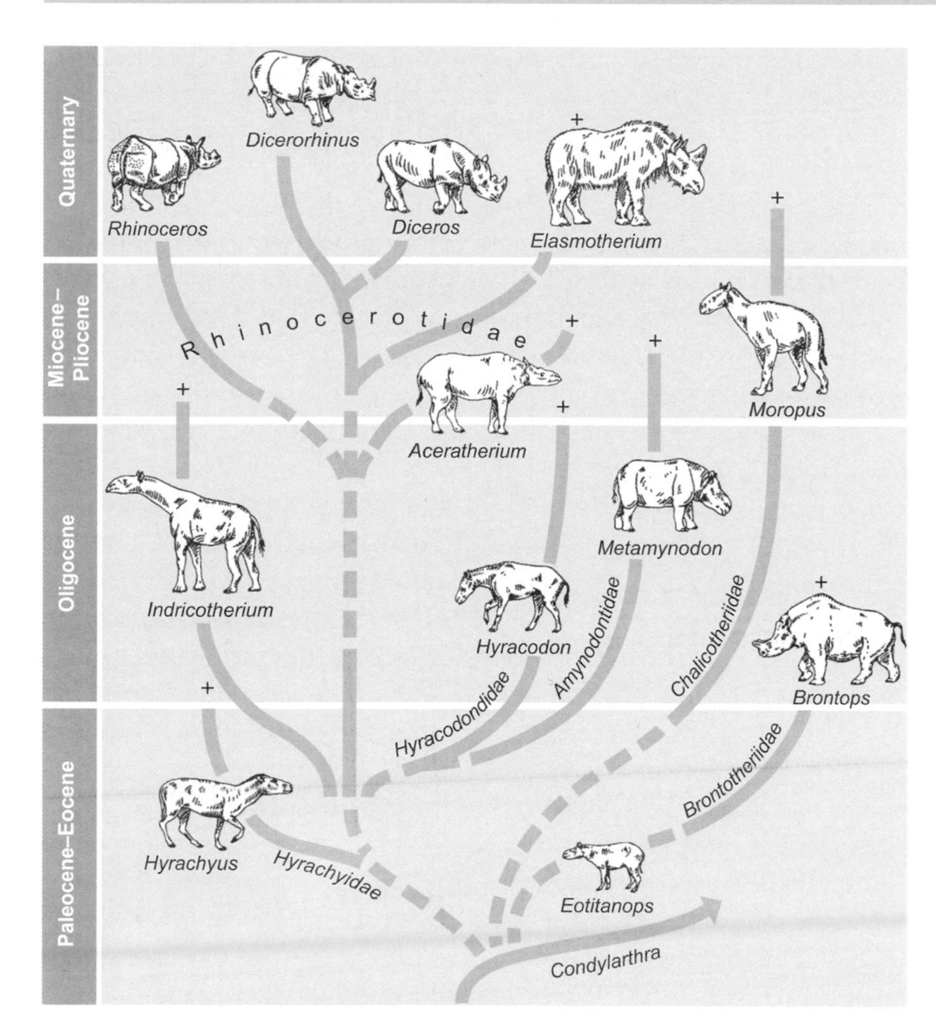

Figure 9.7. Family tree of the rhinos, chalicotheres, and brontotheres.

oped real trunks similar to tapirs. The other genus developed large, massive bodies. These animals, the metamynodonts, turned to an amphibian lifestyle like *Hippopotamus* (Figures 9.3 and 9.7). They lived during the Late Eocene and Early Oligocene of Asia. Some of them emigrated from there to North America. Today, their fossil remains are found in river deposits, so-called Metamynodon channels, in the badlands of South Dakota. At the end of the Early Oligocene, the metamynodonts went extinct in Asia and North America; in Europe they survived in the form of *Cadurcotherium* until the Late Oligocene. The last survivors appeared in Pakistan and Myanmar (the former Burma) as late as the Early Miocene.

The hyracodonts (from Latin *hyrax* = dassie, Greek *odóntes* = teeth) are another group dating back to *Hyrachyus*. Because of their slender limbs they are sometimes called running rhinos. The number of forelimb hooves, already reduced from four to three, is expressed in the name of the earliest genus of this group, *Triplopus* (from Greek *triplóus* = threefold, *pous* = foot). The hyracodonts appeared with several genera in the Late Eocene of North America and Asia (Figure 9.3). Few of them survived into the Oligocene. The best-known genus is *Hyracodon* itself; many of its fossil remains have been discovered in the badlands of South Dakota. These animals, about the size of a Great Dane, were larger than *Mesohippus*, the horse of that time. *Hyracodon* had a somewhat longer skull and stronger and longer neck. With their relatively simple and low-crowned cheek teeth, these animals presumably fed on the leaves and fruits of shrubs and bushes. *Hyracodon* survived into the Late Oligocene. The last representative of the whole group, *Hyracodon* went extinct in North America around 28 million years ago. At the beginning of the Oligocene, *Ronzotherium* appeared for a short time as a member of this group in Europe. Presumably, this genus came from Asia, where the hyracodonts survived in the form of the genera *Ardynia* and *Allacerops* from the Early until the Late Oligocene.

The paraceratheres originated from the genus *Forstercooperia*, which is known from the Middle Eocene of North America as well as China. Since the Oligocene, the genus occurred only in Asia. Their tendency to grow to an enormous size is typical for the paraceratheres. *Forstercooperia* was the size of a cow. *Urtinotherium* from the Early Oligocene of Mongolia reached the size of an elephant. It developed into the genus *Indricotherium*, remains of which have been found in Pakistan, Romania, and the Balkan. During the Late Oligocene, these animals attained a shoulder height of almost 6 meters. Weight estimates range from 15 to 20 tons, up to 35 tons. They were the biggest land mammals of all time (Figure 9.7). These animals were almost the size of the biggest dinosaurs. They could browse in treetops as tall as 7.5 meters. Their skulls attained a maximum length of about 1.5 meters. Among mammals, only the whales grew larger.

Rhinos *sensu stricto*, the family of the Rhinocerotidae, are also considered the descendants of *Hyrachyus*. They appear with the genus *Teletaceras* for the first time in the Middle Eocene of North America and Asia. In the beginning, there were few differences between them and the hyracodonts. Only their upper incisors had developed a cutting function, whereas those in the mandible resembled tiny tusks. With *Trigonias* from the Late Eocene to Early Oligocene, the early rhinos achieved the size of a cow. The skull already displayed the saddle-form depression, which became typical for rhinos in general. Toward the end of the Oligocene, the genus *Diceratherium* developed in North America out of *Subhyracodon*. *Diceratherium* was the first rhino that had horns; as the name indicates, it had two of them. The horns stood next to each other, rather than one behind the other on the tip of the nose as is the case with later rhinos.

The climax for the superfamily of the Rhinocerotoidea was during the Late Miocene. Only in Germany were there no fewer than four different species of these animals living at the same time. Descendants of *Diceratherium* that immigrated into Europe diversified into different groups, such as the Teleoceratini, the Rhinocerotini, and the Aceratheriini, whose horns were reduced in favor of cutting, daggerlike lateral incisors. Unlike the Oligocene rhinos, however, these were in the lower, not upper, jaw. Throughout evolutionary development, constructional conditions are tested, then further developed or rejected.

The evolution of the Miocene rhinos started with a form that already had enlarged lower canines and relatively small horns. Their subsequent development was characterized by increased size in one or the other of these two features. If the horns increased, the lower canines decreased, and vice versa. The limbs also changed. Four-hoofed forelimbs and three-hoofed hind limbs were present in the Early to Middle Eocene *Hyrachyus*. During the Early Miocene the Teleoceratini developed short, but very strong, extremities called graviportal (from Latin *gravis* = heavy, *portare* = to carry). This development culminated in the Late Miocene *Brachypotherium goldfussi*, a typical representative of the so-called *Hipparion* fauna of Europe. Some hornless Aceratheriini also shortened their limbs during the Late Miocene, although to a much lesser degree. This was combined with an increase in body weight. Additionally, several genera of both groups—Teleoceratini and Aceratheriini—reduced the number of toes and hooves in the forelimb from four to three independently of each other. Among the Aceratheriini, a member of this kind was the species *Aceratherium incisivum*, a rather common companion of the Late Miocene horse *Hippotherium primigenium*.

In addition to changes in horns, canines, incisors, and limbs, important changes within the skulls occurred on the way to modern rhinos. While in early rhinos palate and cheek teeth lay in the same plane as the skull base, an obtuse angle between them developed during the Miocene. This occurred together with a deepening of the nasal incision and with a shortening of the nasal bones.

The group of modern rhinos, the Rhinocerotini, appeared in Europe for the first time during the Early Miocene with the strong-horned genus *Lartetotherium*. This genus went extinct during the Late Miocene. At the same time, a series of immigrants appeared in Europe, among which were the Rhinocerotina from Asia and the Dicerotina from Africa. To the first group belongs *Dihoplus schleiermacheri*, the biggest rhino of the whole Miocene. Two massive horns, one behind the other, and a long, broad skull characterized this species, which was a typical member of the central European Hippotherium faunas.

At the end of the Tertiary, rhinos even evolved to survive in the cold steppes of ice age Europe. The woolly rhino (*Coelodonta antiquitatis*), with its thick red brown fur, was a characteristic faunal element of the final ice age. The fur has been preserved in some carcasses found in the Siberian permafrost. Our ancestors painted images of the impressive animal on the walls of caves in southern France and northern Spain during the Late Pleistocene (Figure 9.8). Through the eyes of our ancestors we have an idea of what these animals looked like. The genus *Elasmotherium*, which had developed at the beginning of the ice age in China, reached the size of elephants and had only one gigantic horn on the nasals. Later on, they occurred mainly in Siberia, where they obviously grazed on the so-

Figure 9.8. Woolly rhino (*Coelodonta antiquitatis*), 13,000- to 24,000-year-old wall painting of the late ice age. The painting belongs to the so-called frieze of the three rhinos in the cave of Rouffignac (southern France). Note the thick, furlike hairs on the lower side of the body.

called mammoth steppe of the ice age. This is indicated by high-crowned cheek teeth with an extremely folded occlusal pattern, similar to grazing horses. Like many other gigantic mammals, *Coelodonta* and *Elasmotherium* went extinct at the end of the ice age, around 10,000 years ago.

There were important families among the perissodactyls that did not survive into the present time (in contrast to the horses, tapirs, and rhinos). Among the most amazing of these are the claw-bearing ungulates, or chalicotheres. Fossil remains of this group provided Georges Cuvier—the founder of vertebrate paleontology—with one of his most difficult problems. Cuvier had formulated the so-called rule of correlation: "Each organism is a sum of details, a closed system of its own, the parts of which correspond with each other, so that each of them taken for itself indicates all the others." For example, cats are characterized not only by their typical carnassials but also by retractable claws. Similarly, modern horses are characterized not only by their extremely high-crowned cheek teeth but also by their solid hoofedness. Paleontologists—the explorers of prehistoric animals—still apply that rule when they work on isolated teeth. They think about the complete skeleton and the complete animal. But no rule is without exceptions. Cuvier was the first to fall victim to the traps of his own rule. One day, large phalanges split at their ends were brought to him for identification. Cuvier did not hesitate to identify them—faithfully trusting in his rule of correlation—as the fossil claws of a giant pangolin (scaly anteater). At the same time, he determined isolated teeth from the same locality were those of a large perissodactyl. After Cuvier died the discovery in France of a whole skeleton, with teeth and claws of just that kind, created a sensation. Obviously, the teeth and claws must have belonged to the same species, which was given the name *Chalicotherium grande*. These animals had a strange construction (Figure 9.9). While their distal phalanges resembled those of pangolins and large sloths, the proportions of their body, the long, strong arms and short legs, were more like that of a giant gorilla. Moreover, the long stretched-out skull displayed cheek teeth that were quite similar to those of rhinos. What can one say about an animal that has no extant analogue? Today, experts agree that the whole group of chalicotheres belong to the perissodactyls. It is still not clear, however, if within the Perissodactyla, the closest relationships exist with rhinos, horses, or tapirs. The next nearest relatives of the chalicotheres were the lophiodonts, which have also been extinct for a long time. Less doubt exists about their way of life than about their systematic posi-

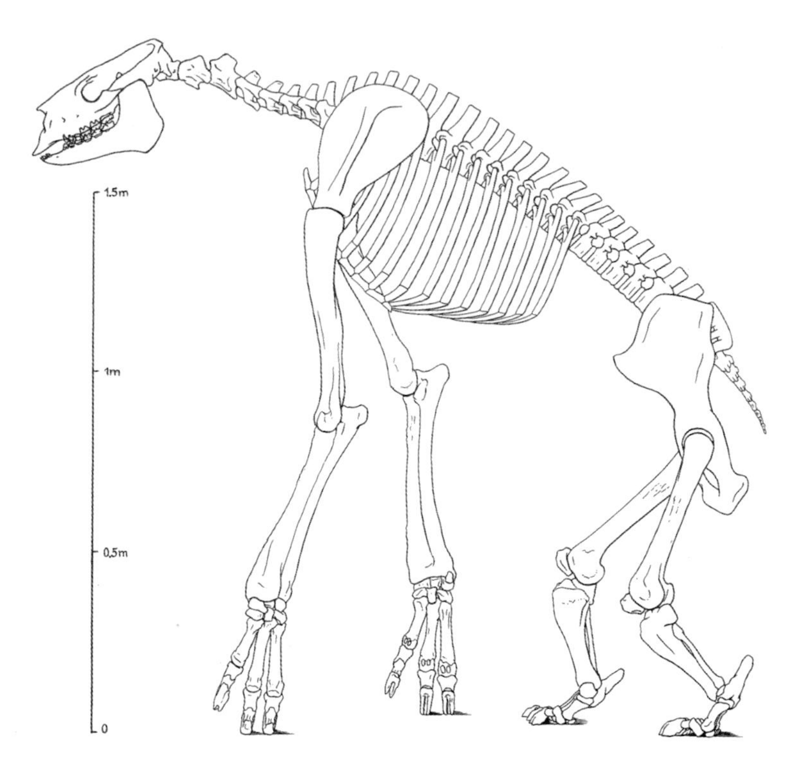

Figure 9.9. Skeletal reconstruction of *Chalicotherium grande*, based on bones from the Děvinská Nová Ves (formerly Neudorf an der March) in Slovakia.

tion (Figure 9.10). Their low-crowned dentition resembles that of mammals that feed on leaves and fruit. Their long arms with hook-like hands may have been used to pull branches down in order to obtain such food. Presumably, these animals moved not very differently from gorillas today.

Chalicotheres are rare at sites with tertiary mammals. There are, however, exceptions where such finds are comparatively frequent, as at Neudorf an der March (now in the Slovakian Republic and known as Děvinská Nová Ves, and at Eppelsheim in the Mainz Basin (Germany). At Děvinská Nová Ves, chalicotheres are so common that Helmut Zapfe, the former mammalian paleontologist from Vienna (Austria), was able to prepare a monograph on *Chalicotherium grande* in which he completely reconstructed the skeleton and bauplan of that species (Figures 9.9 and 9.10). With regard to the phylogenetic development of the chalicotheres, the oldest representatives of this group are known in the forms of *Eomoropus* and *Grangeria* from the Middle Eocene of Asia and North America, respectively. Alleged finds of dawn horses from the Early Eocene of East Asia, such as *Propachynolophus hengyangensis, Propalaeotherium sinense,* and *Gobihippus gabuniai,* may also belong to this group (see Chapter 8). During the Oligocene, only one genus, *Schizotherium,* occurred in

Europe and Asia. However, during the Miocene, the two groups inhabited the same area, the genuine chalicotheres represented by *Chalicotherium*; and the schizotheres, by the genus *Moropus* (Figure 9.3). Originating in Eurasia, the schizotheres ventured into North America during the Late Oligocene and into Africa during the Late Miocene. There, they survived together with the horse *Stylohipparion* into the ice age. The chalicotheres succeeded in China. After that, the whole family, which was successful for millions of years, went extinct. We do not know why. Perhaps it was the extreme Pleistocene oscillations of climate and vegetation that caused their extinction.

A group that was almost entirely restricted to the Eocene and Early Oligocene of North America and Asia were the thunder beasts, or brontotheres (from Greek *bronte* = thunder, *theríon* = mammal). These were massively built perissodactyls (Figures 9.3 and 9.11). Toward the end of their evolution, with paired horns on their noses, they filled the role of the later, more common rhinos. With shoulder heights of up to 2.5 meters for *Brontotherium*, they developed remarkable body sizes that clearly surpassed those of modern rhinos. Remains of a genus called *Brachydiastematherium* from Romania, Bulgaria, and Hungary prove that single members of this group crossed the European border, although they did not become indige-

Figure 9.10. Artist's rendering of *Chalicotherium grande* in its natural habitat. On exhibit at the Naturhistorisches Museum in Basel (Switzerland).

Figure 9.11. Evolution of the brontotheres during the Eocene and Oligocene.

nous on this continent. We do not know what was responsible for this. Perhaps they did not find the right food; perhaps the climatic conditions were not right for them.

The origin of the thunder beasts is unknown. *Lambdotherium* and *Eotitanops*, the earliest representatives, appeared during the Early Eocene in North America without showing any hint as to their ancestors. Some have tried to trace the whole group back to *Hyraco-*

therium, while others have pointed to similarities between *Lambdotherium* and the European palaeotheres. The origin of this group, however, is still uncertain.

Among the extinct perissodactyls, the palaeotheres (from Greek *palaios* = old, *theríon* = mammal) were most similar to horses. These include representatives of such genera as *Palaeotherium, Pseudopalaeotherium, Paloplotherium, Plagiolophus,* and *Leptolophus.* With the exception of a few finds from the Balkans, almost all are known exclusively from west and southwest Europe. Some special developments like *Cantrabrotherium* and *Franzenium* occurred only on the Iberian Peninsula. *Bepitherium,* from the Early Eocene of the Ebro Basin in northeast Spain, has been identified as the earliest member of the whole group, pointing to the group's geographic origin.

It is in the structure of their dentition that the palaeotheres, or tapir horses, resemble genuine horses to such an extent that even internationally renowned specialists are convinced of their very close relationship. There are, however, fundamental differences that preclude the possibility of deriving the palaeotheres from *Hyracotherium,* the phylogenetic root of all equids. One such differentiating characteristic exists in the structure of the pelvis, which still resembles that of archaic ungulates among the Condylarthra. They were clearly more primitive than *Hyracotherium,* whose pelvis already resembled modern horses. The same holds true for the metatarsals, which—contrasted with all equids including *Hyracotherium*—are shorter, not longer, than the accompanying metacarpals. The only member of that group that is represented by a complete skeleton—*Palaeotherium magnum* from the plaster quarries of former Montmartre and Mormoiron in southern France—shows that the physics of these animals was different from that of all equids. The front part of the body of *Palaeotherium magnum* was considerably higher than its hind part, and the neck was proportionally much longer than that of all contemporary horses (Figures 9.3 and 9.12). These animals were "anteriorly overbuilt." In this respect, they resembled giraffes, although they were definitely more ponderous, at least as far as the members of the genus *Palaeotherium* are concerned. For all these reasons, it is evident that the palaeotheres do not go back to *Hyracotherium.* Consequently, they are not equids, or hence horses. They have to be considered as their own independent family. Their stretched-out forelimbs and neck together with the structure of their dentition indicate that these animals were browsers, and hence folivorous. On the other hand, some of the genera—such as *Plagiolophus, Leptolophus,* and *Cantabrotherium*—reacted to the increas-

Figure 9.12. Artist's rendering of palaeotheres (*Palaeotherium magnum*) in their natural habitat.

ingly dry climatic conditions in Europe by developing high-crowned cheek teeth with cement deposits inside and outside that were especially suited for processing hard instead of soft nutrition. At the beginning of the Late Eocene, palaeotheres like *Palaeotherium* and *Paloplotherium* appeared in central and western Europe next to the early representatives of genuine horses, such as *Propalaeotherium, Eurohippus, Lophiotherium,* and *Anchilophus*. Under the increasingly dry climatic conditions during the Late Eocene, palaeotheres more and more replaced the equids, and they survived into the Early Oligocene, while horses went extinct in Europe with *Anchilophus radegondensis*—the last representative—during the Late Eocene, about 35 million years ago.

The palaeotheres pursued two fundamentally different directions in the development of their dentitions. Some genera—such as *Plagiolophus, Paloplotherium,* and *Leptolophus*—emphasized the development of their hind cheek teeth, the permanent molars, which became big and remarkably high-crowned while the premolars remained small, low-crowned, and rather primitive in their occlusal pattern. The dental development of the genera *Palaeotherium, Cantabrotherium,* and *Pseudopalaeotherium* enlarged not only the molars but also the premolars. These also became more similar to the molars. This process, which also occurred in the horses, is called molarization of the premolars. The cheek teeth remained, however, rather low crowned.

Although palaeotheres are not horses, they are close relatives of them. In South America, which was completely isolated at that time, true pseudo horses developed. Within the order Litopterna (from Greek *litós* = straight, *ptérna* = heel), which dates back to Condylarthra or primitive ungulates, these included the Proterotheriidae (from Greek *próteros* = anterior, *theríon* = mammal) with genera such as *Diadiaphorus* and *Thoatherium.* Also included are the Notohippidae, which are within the order Notoungulata (from Greek *nótos* = south, Latin *ungula* = hoof; hence, the southern ungulates). Typical dentitions of grazers—very high-crowned molars, tooth cement, and highly complex wear patterns—developed in the Notohippidae during the Oligocene, hence much earlier than in horses. On the other hand, the Proterotheriidae retained the typical low-crowned and comparatively simple cheek teeth of browsers throughout their whole evolution. The skull, body, and limbs, however, became very similar to horses. In the course of its development at the beginning of the Miocene (almost 20 million years earlier than horses), the genus *Thoatherium* had already achieved a degree of solid hoofedness. This solid hoofedness even surpassed the reduction of the side toes in modern *Equus* (Figures 9.13 and 9.14). Amazingly, these pseudo horses were not very successful and went extinct long before true horses even reached South America. Perhaps the reason for this was that they were solid-hoofed browsers and not grazers like true horses, that is, an increasing discrepancy developed between the construc-

Figure 9.13. Sketch of what the South America pseudo horse *Thoatherium* may have looked like.

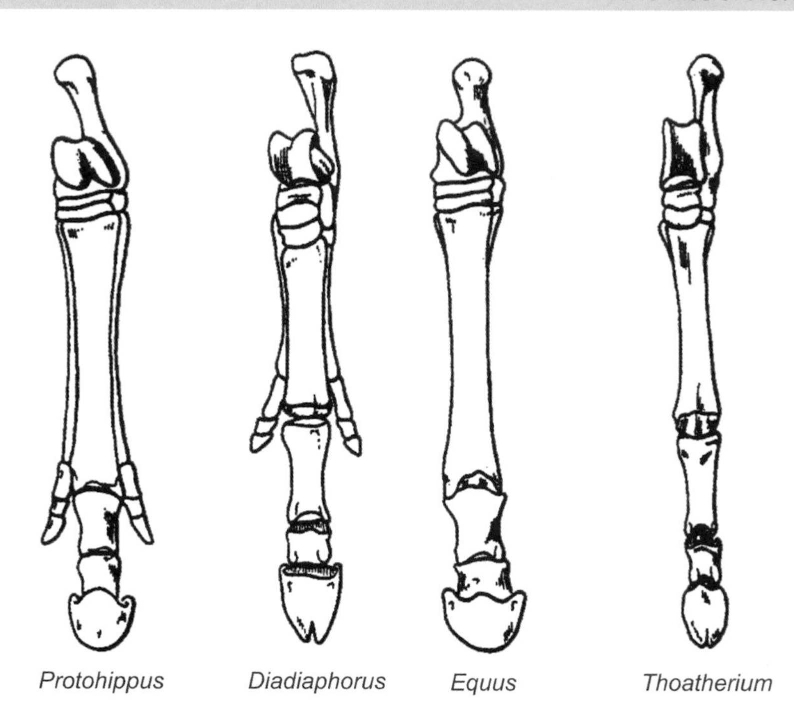

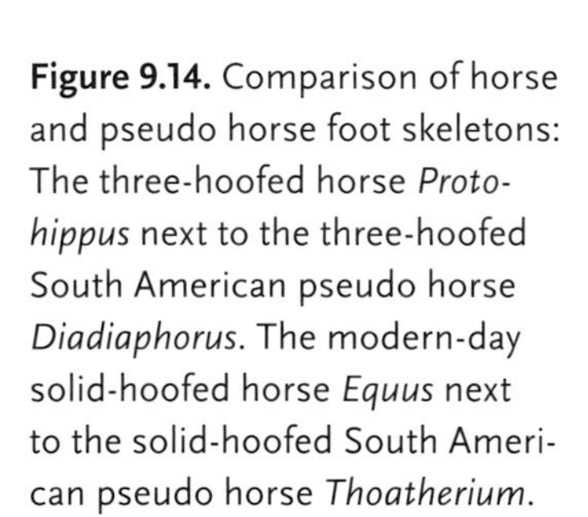

Figure 9.14. Comparison of horse and pseudo horse foot skeletons: The three-hoofed horse *Protohippus* next to the three-hoofed South American pseudo horse *Diadiaphorus*. The modern-day solid-hoofed horse *Equus* next to the solid-hoofed South American pseudo horse *Thoatherium*.

tional requirements of food uptake and the environment on one side and fast locomotion on the other.

It is of interest that in all these cases, whether we are dealing with South American notohippids and proterotherids, European palaeotheres, or North American horses, quite similar developments took place as far as constructional preconditions and environmental limitations allowed.

CHAPTER TEN

The Ice Age and the Roots of Modern Horses

THE IDEA OF A LONG-AGO ICE AGE when hordes of Neanderthals roamed the tundra between polar ice sheets in the north and icy mountain ranges in the south was once quite popular. But we now know that such a scenario is inaccurate. Not only did Neanderthals appear rather late in human evolutionary history, but they also left it rather early, replaced by our own ancestors—modern *Homo sapiens*—when they emigrated from Africa. And there was not one single ice age but rather up to 16 separate ice ages, or glacials, interspersed with a variety of interglacials, or intercalated warm times, of varying lengths (Figure 10.1). Altogether, the ice ages, or glacial periods, comprised about twice the amount of time as the intercalated warm periods. During those interglacials the average annual temperatures were similar to those of today or even a few degrees centigrade higher. Climates also fluctuated within the individual glacials and interglacials. The 10,000 years from the end of the last glacial period to the present were a time of extraordinarily stable climatic conditions. This was probably one of the most decisive preconditions for the emergence and development of modern human societies. From the point of view of a geologist or climatologist, we are living in an interglacial. In order to set people's minds at ease, one should realize that the next glacial period is probably many thousands of years away. How is it possible to say that? What is the basis for such a prediction?

Although the myriad factors affecting climate change have not all been identified, one thing is clear: Astronomic cycles and their interferences have had a definite impact on climate during the

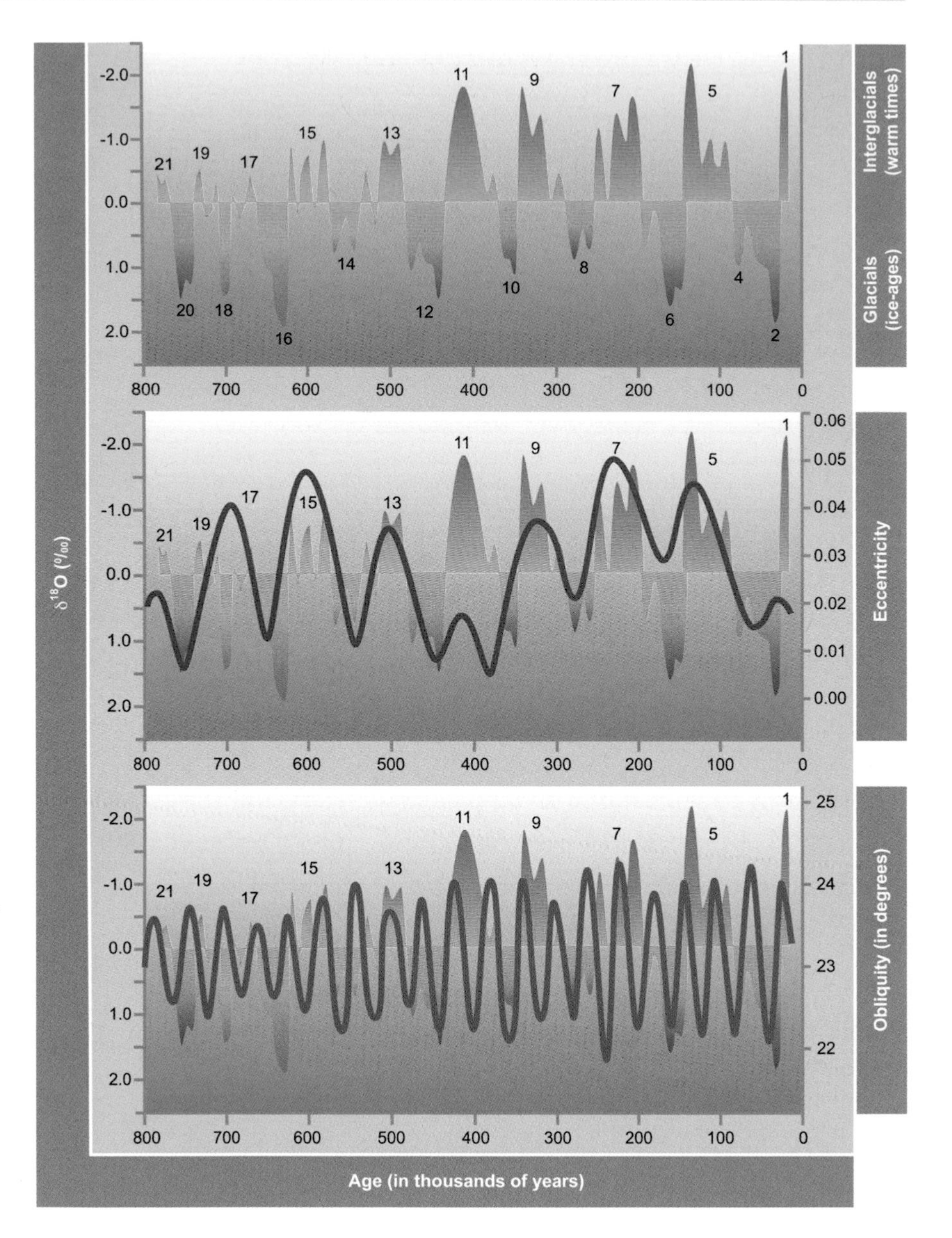

Figure 10.1. The ice age (Pleistocene) consists of glacials (even numbers) and interglacials (odd numbers). It is possible to reconstruct the temporal development of paleotemperature (*top*) from the content of oxygen isotope $\delta^{18}O$ in foraminifera shales, which indicates the respective temperature. These correspond to the earth's orbit—the precessions of the eccentricity (*middle*: red curve) and of the obliquity (*bottom*: gray curve); both change periodically at different rates. (See also Figure 10.2.) The paleotemperature curve corresponds amazingly well with curves of paleoclimatic development, which were the result of analyses of core drillings from the ice shields of Greenland and the Antarctic. The graph shows the development of paleoclimate during the last 800,000 years of the Pleistocene.

earth's history and will continue to do so in the future (Figure 10.2). One cycle is the elliptical form of this orbit (eccentricity). During one circulation (i.e., one year), the earth is at one time closer to the sun, at another farther from it. Another of these cycles is the obliquity of the earth's axis of rotation as it revolves around the sun. The result is cyclical annual seasons that occur at opposite times in the Northern and the Southern Hemispheres. In other words, during Northern summertime there is winter in the Southern Hemisphere and vice versa.

These circumstances, however, only explain the annual seasons.

With regard to long-term climatic changes, the important fact is that the plane of the earth's orbit around the sun is wobbling at a cycle of 19,000 years, while the obliquity of the earth's axis relative to the orbit swings at a cycle of 23,000 years. Simply put, the earth doesn't turn but rather staggers around the sun. Based on the interference of these two rhythmic oscillations—which are also called the precession of the orbit, or the obliquity of the earth's axis—it is possible to calculate a long-term curve of climatic development that corresponds to what occurred in the past. Deep drillings into the ice sheets of the Antarctic and of Greenland made it possible to analyze climatic developments of the past. The "eternal ice" of the polar regions is not, in fact, eternal; changes in annual precipitations resulted in its development. The magnitude and rhythm of these precipitations change to some extent each year. Using drill cores from deep inside the earth, this rhythm can be analyzed and dated. The ice sheets of the polar region represent a frozen calendar of climatic development during the ice ages, or the Pleistocene, that is, the last 2 million years of the earth's history. The constant astronomic parameters are also used to determine climatic developments in the future (Figure 10.3). Nonetheless, such predictions are not fail-proof. Variables such as fluctuations of the energy radiation from the sun and the human impact on the earth's atmosphere (the greenhouse effect) also play a role. Moreover, calculations concerning the volume of ice in the Northern Hemisphere are in dispute as estimates of the future content of carbon dioxide in the atmo-

Figure 10.2. The astronomic parameters that are decisive for the climatic development on earth. *Clockwise from top left*: eccentricity, the elliptical orbit of the earth around the sun; obliquity, the inclination of the earth's axis of rotation; the wobbling (precession) of the orbit's plane in a cycle of 19,000 years; the swaying (precession) of the earth's axis of rotation in a cycle of 23,000 years.

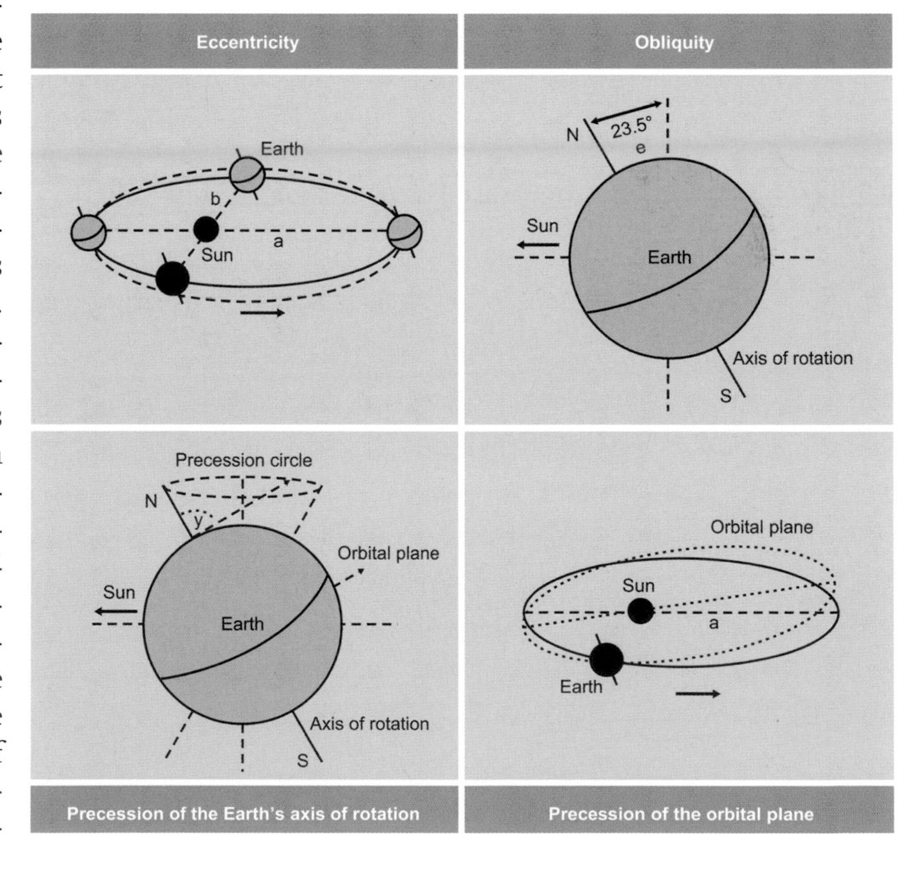

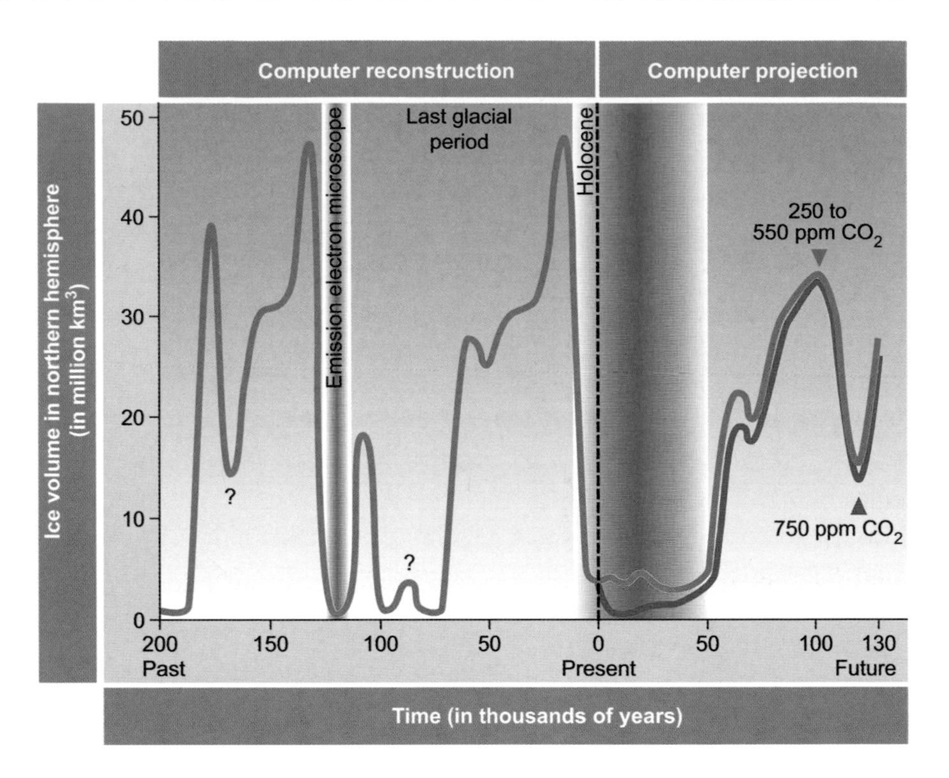

Figure 10.3. Development of the earth's climate based on astronomic parameters. (See also Figures 10.1 and 10.2.) The estimates of the ice volume in the northern hemisphere are uncertain for some time intervals. The atmospheric content of carbon dioxide in the future, estimated at 750 ppm, is also questionable.

sphere. So how did climate changes influence the evolution of the horses?

During the Pliocene, around 4 million years ago, horses developed in North America out of the genus *Dinohippus*. The most conspicuous morphological difference between *Dinohippus* and its phylogenetic descendants was that the depressions in the skull in front of the eye sockets—the preorbital fossae that are so typical for Miocene horses—vanished completely (Figures 10.4 and 10.5). Neither their function nor the reason for their disappearance is clearly understood. Possibly these fossae held glands whose secretions served to mark habitats. Such glands could have been important during times of high diversification—as was the case during the Miocene—and less important during the Pleistocene, when the number of horse genera decreased enormously (see Figure 6.19). Although the morphological differentiation of horses was high during the ice age—from America alone there are more than 50 supposed species of *Equus*—molecular genetic analyses show that there were really only a few "genuine" species at that time. These were probably mostly polymorphic. Thus paleontological investigations seem to have generated far more species than were actually present.

Several genera are recognized among the direct descendants of *Dinohippus*. For example, *Plesippus* (from the Greek *plesios* = close, and *(h)ippos* = horse) appeared more than 3 million years ago. *Allohippus* (from Greek *allos* = different) came a short time later and is distinguished from *Plesippus* by its relatively deep nasal incision. Both differ from *Equus* by their comparatively short brain case (Figures 10.5 and 10.6). *Allohippus* as well as *Plesippus* first turned up in China about 2.5 million years ago, soon after their emergence. *Allohippus* appeared a little later in Europe, and still later in East Africa (Figures 10.7 and 10.8). The oldest skull of *Equus* was found in 2-million-year-old layers exposed in the Anza Borrego Desert in Southern California. In the beginning, the species was erroneously named *simplicidens,* which actually belongs to the genus *Plesippus.* From the modern viewpoint, the genus *Equus* comprises all living equids—horses, zebras, asses, and half-asses—and all Pleistocene horse species with the exceptions of *Allohippus* and the South American *Hippidion* and its relatives (Figure 10.8).

The beginning of Pleistocene is defined by the so-called Olduvai event, which was a reversal of the earth's magnetic field around 1.8 million years ago. The climatic change, however, had occurred much earlier, around 2.6 million years ago. So there is a temporal congruence with the emergence of the genus *Equus.* It cannot be said with certainty whether there was also a causal connection between the two, although the increase in brain capacity could have made it easier to adapt to the fast-changing conditions of life during the Pleistocene. It would also explain the disappearance of *Plesippus* and *Allohippus* at the end of the Pliocene or beginning of the Pleistocene; both genera were characterized by their relatively small brain as compared with *Equus.*

The origin of modern horses, which includes not only *Equus caballus* but also zebras, the half-asses, and the asses, dates back to the Middle Pleistocene. Before that, the tracks of evolution are not visible. The oldest European species that comes close to the origin of modern horses is *Equus mosbachensis* (Figure 10.9). Its remains occur all over central Europe in Middle Pleistocene deposits. This

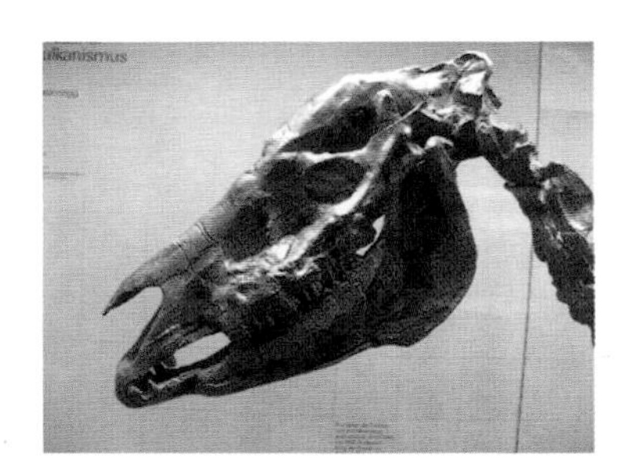

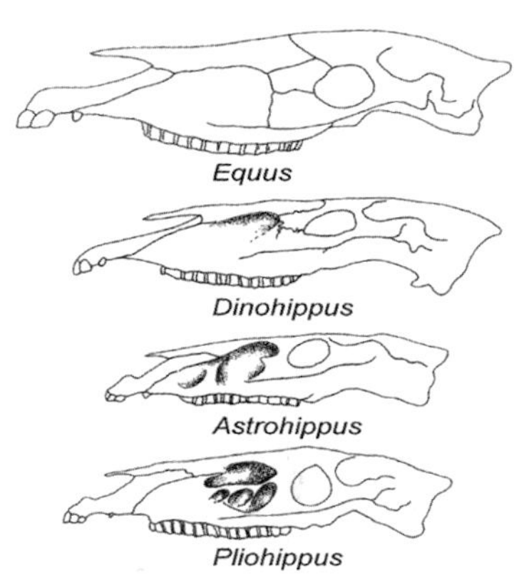

Figure 10.4. *(Top)* Depressions situated in front of the eye sockets (preorbital fossae) on the cast of a skull of Late Miocene *Hippotherium primigenium* that was found in the vicinity of Höwenegg (Germany). The cast is on exhibit at the Staatliches Museum für Naturkunde in Stuttgart. **Figure 10.5.** *(Bottom)* Preorbital fossae on the skulls of *Pliohippus, Astrohippus, Dinohippus,* and *Equus.*

Figure 10.6. Side views of *Allohippus, Hippidion,* and *Equus* skulls.

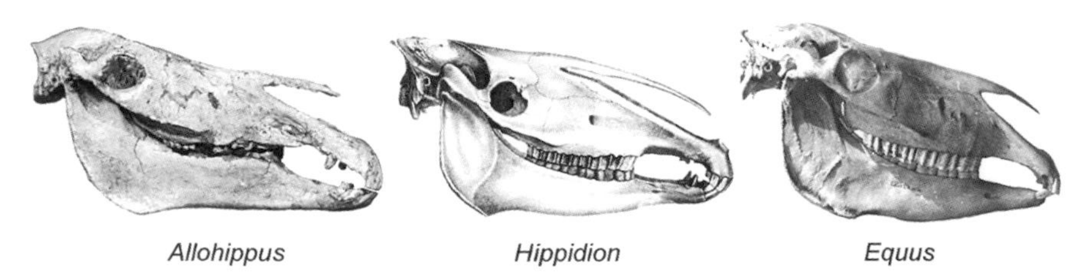

Figure 10.7. About 2.5 million years BP, *Allohippus stenonis* appears in Europe as the earliest ancestor of horses *sensu stricto* (subfamily Equinae). The skeleton, from Senèze on the French central plateau, is on exhibit at the Naturhistorisches Museum in Basel (Switzerland).

species is not derived from the Early to Middle Pleistocene Eurasian horse species such as *Equus suessenbornensis*, *E. coliemensis*, *E. nalaikhaensis*, or *E. granatensis* (the oldest *Equus* species from Asia). More likely, *Equus mosbachensis* is an immigrant originating from the North American *Equus scotti*, whose phylogenetic root is also obscure (Figure 10.8).

Concerning the direct origin of modern *Equus caballus*, most specialists agree that the wild-living horses at the end of the Pleistocene of Eurasia—known as *Equus gallicus* or *Equus germanicus*—go back to Middle Pleistocene *E. mosbachensis*. In addition to fossils, cave paintings, dating from the end of the last glacial period (Figures 10.10 and 10.11; see also Figure 2.9) in southern France, northern Spain, and the southern Urals (Russia), demonstrate that the European "cave horses" were clearly different from the tarpan (*Equus "ferus"*; see Figure 2.6d), and from the Przewalski's horse (*Equus "przewalskii"*; see Figure 2.6c). They are differentiated from the tarpan by their short skulls and from the Przewalski's horse by the outline of their skulls and their relatively short metapodials (Figure 10.11). The mechanisms of if and how the modern domestic horses developed from the "cave horses" or from the tarpan are uncertain because of the complex mixture of the various races. At the same time in southern Europe there was another type of horse (*Equus antunesi*), which was characterized by a long facial part of the skull and slender limbs. This too could be an ancestor of modern horses. The typical DNA of the Przewalski's horses is not present in any of the races of modern domestic horses, indicating that the Przewalski's horses represent a local development from East Asian steppes and semideserts that is not directly related to modern domestic horses. This idea is further strengthened by their long limbs and slender hooves, which are especially suited for fast locomotion (Figure 10.11). It has been proven that the few so-called wild horses of today—such as the Duelmen horses and the Exmoor ponies (see Figure 2.6a, b)—have been drastically changed by man's breeding strategies and that they do not have much in common with the original wild horses. And the North American mustangs are simply domestic horses that ran wild (feral).

Figure 10.8. Overview and phylogenetic relationships of horses occurring from the Pliocene up until postglacial times in Africa, Eurasia, and North and South America.

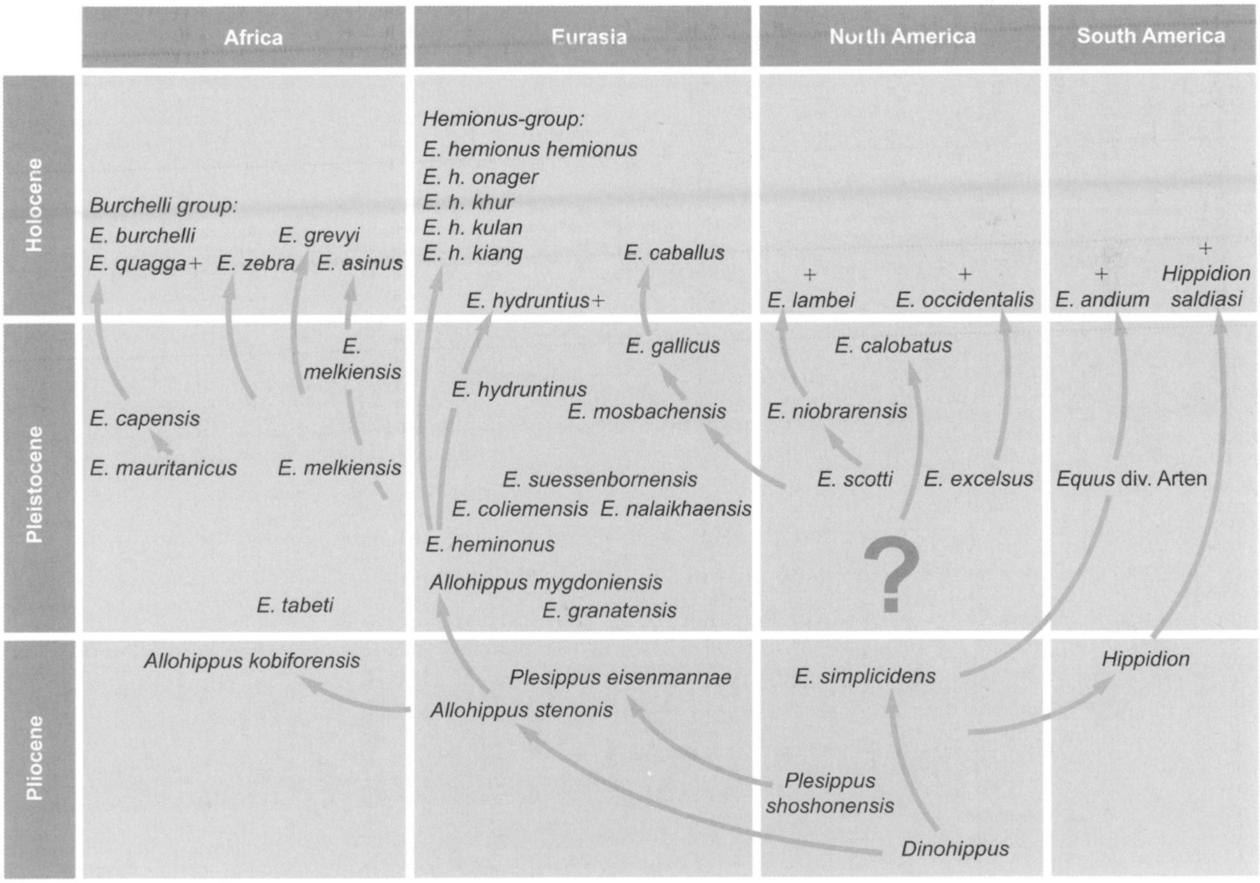

	Africa	Eurasia	North America	South America
Holocene	*Burchelli group:* *E. burchelli*, *E. grevyi*, *E. quagga*+, *E. zebra*, *E. asinus*	*Hemionus-group:* *E. hemionus hemionus*, *E. h. onager*, *E. h. khur*, *E. h. kulan*, *E. h. kiang*, *E. caballus*, *E. hydruntius*+	+ *E. lambei*, + *E. occidentalis*	+ *E. andium*, + *Hippidion saldiasi*
Pleistocene	*E. melkiensis*, *E. capensis*, *E. mauritanicus*, *E. melkiensis*, *E. tabeti*	*E. gallicus*, *E. hydruntinus*, *E. mosbachensis*, *E. suessenbornensis*, *E. coliemensis*, *E. nalaikhaensis*, *E. heminonus*, *Allohippus mygdoniensis*, *E. granatensis*	*E. calobatus*, *E. niobrarensis*, *E. scotti*, *E. excelsus*, ?	*Equus* div. Arten
Pliocene	*Allohippus kobiforensis*	*Plesippus eisenmannae*, *Allohippus stenonis*	*E. simplicidens*, *Plesippus shoshonensis*, *Dinohippus*	*Hippidion*

Figure 10.9. *Equus mosbachensis* of the Middle Pleistocene in Europe was probably the ancestor of modern *Equus caballus*. The skeleton, which is about half a million years old, was constructed from fossils found in Rhine River sediment near Mosbach in the vicinity of Wiesbaden (Germany). It is on exhibit at the Naturhistorisches Museum in Mainz.

The origin of zebras and wild asses (Figure 10.8) can be traced back to a group of Middle Pleistocene steppe zebras. The 700,000-year-old *Equus mauretanicus* of North Africa and the somewhat younger *E. capensis* of South Africa show for the first time differentiations in the direction of zebras. Similarly, the Ethiopian wild asses—from which the modern asses (*E. asinus*) are derived—go back to *E. melkiensis*, a species known from the Middle to the Late Pleistocene of North Africa. The half-asses of the steppes and deserts of the Near East and Central Asia can be traced back to roots that are around 1 million years old. *E. hydruntinus*, the so-called European ass of the last glacial period, is in fact a side branch of the half-asses that went extinct during early postglacial times.

The relationship between humans and horses during the Pleistocene—initially, horses viewed as prey, then, close to the end of the last glacial period, as highly honored cult figures—is discussed in

Chapter 2. Here we deal with a hotly debated issue: the case of the extinction of the species *E. lambei*, *E. calobatus*, and *E. occidentalis* in North America shortly after the end of the last glacial period. What was the reason for the sudden disappearance of horses in North America after millions and millions of years, and why, at the same

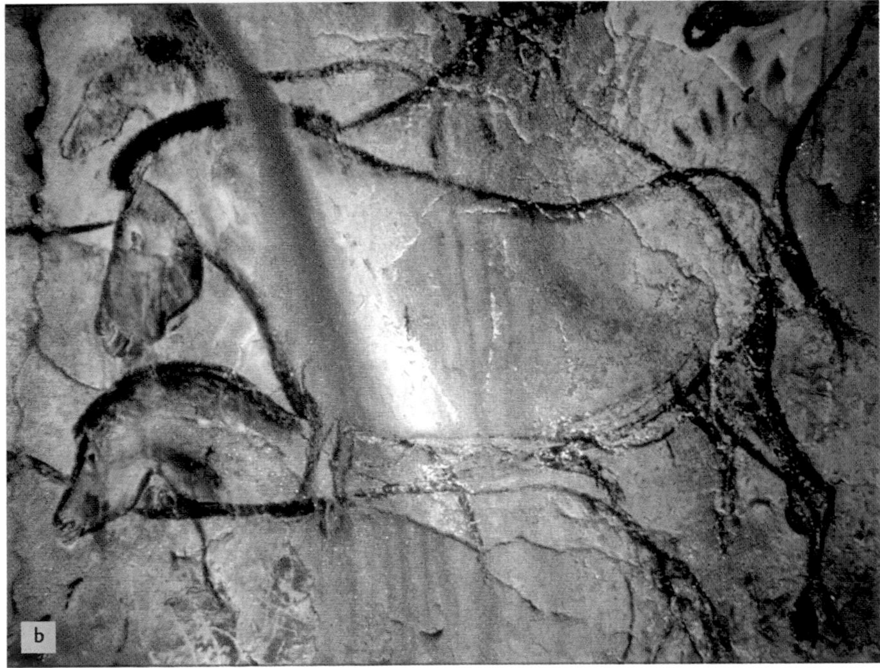

Figs. 10.10. Paintings, about 32,000 years old, of Late Pleistocene horses on the walls of Grotte Chauvet, in the Ardeche Canyon (southern France): (a) The horse heads are realistic and powerful; a woolly rhino (*Coelodonta antiquitatis*) is depicted in the bottom left corner; (b) a thin streak of calcite colored yellow by iron hydroxide testifies to the age and authenticity of the drawing.

areas of that continent, where they survived until the middle ice age in the form of the genus *Stylohipparion*. There, they must have met early members of our own species, as the fossil human footprints from Laetoli in Tanzania suggest. The development of *Hipparion* into *Equus* did not occur. With *Astrohippus*, *Pliohippus*, and *Dinohippus* horses achieved for the first time the ability to stand on one hoof per limb, and even to run. This achievement, which occurred during the Pliocene, included three separate parallel lineages. All three genera go back to *Merychippus* of the Late Miocene. *Astrohippus* and *Pliohippus* soon went extinct. From *Dinohippus* a lineage branched off and led to *Hippidion* and its relatives in South America. An extremely deep nasal incision and short feet characterized these horses. The genus survived in Patagonia until long after the ice age.

Another branch bifurcating from *Dinohippus* during the Pliocene, 2 to 3 million years ago, was the genus *Equus*. From North America, it migrated by way of Asia into Europe and Africa, and somewhat later than *Hippidion* into South America, where it developed with *Equus (Amerhippus) andium*, a species with remarkably short feet that was particularly suited for living in the high mountains of the Andes. In the Old World, the genus *Equus* soon split into branches that led to the asses and half-asses, the zebras, and modern *Equus caballus*, the domestic horse. It is ironic that soon after the ice age *Equus* went extinct in North America where horses had successfully evolved for over 55 million years. The Spanish invaders brought the genus back to its homeland on board their caravels. But this species, *Equus caballus*, had developed in Eurasia. Why *Equus* was able to survive in Eurasia but not in North America has not been resolved. Probably, the fundamental changes in climate and vegetation at the end of the ice age played a decisive role. Presumably, these changes were connected with epidemic diseases that were restricted to North America's horse species. At that time, human beings had neither the weapons nor the population density to wipe out large herds of fast-running horses.

When we look at the evolution of horses as fossils document it (Figure 11.1), we see—in spite of any gaps in detail—a movie-like image of the evolution of animals that were finally connected to humans by at least 4,000 years of coevolution. At various times, horses have migrated as far as land bridges and their body construction would allow. And at all times their evolutionary development—as far as can be determined—followed economic principles in the sense of improving the energy balance. On the other hand, the extent of diversification was determined by the number of different ways in which it was possible to survive—in other words, on the available biota and the various strategies with which one might use them. In that respect the horses' family tree presents a fascinating picture of phylogenetic development. Horses are and remain *the* classic example of evolution.

Fossil Horses in Cyberspace

This is a virtual museum exhibit developed by the Florida Museum of Natural History and partially funded by a grant from the National Science Foundation. The virtual exhibit is easy to navigate, and its content is easy to understand. Exhibits include a gallery of fossil horses, a discussion of names, stratigraphy, geological time scales, and Amazing Feets: Tales Told by Toes (a discussion of comparative and functional anatomy). It is a pity that the exhibit is restricted to the North American fossil record.

Internet: www.flmnh.ufl.edu/fhc/

Darmstadt, Germany

The Hessisches Landesmuseum (Country Museum) in Darmstadt was founded in the late eighteenth century as the natural history collection of the former grand duke of Hessen. By 1817 it had already received rich collections of fossil mammals, including teeth and bones of *Hippotherium primigenium* from the famous Dinotheriensande of Rheinhessen (locality of Eppelsheim). In 1912, it became the first, and was for many years the only, institute given the right to collect the fossils from Messel. The museum's collection of Messel fossils dating to the mining days at Messel includes multiple skeletons of the dawn horses *Propalaeotherium hassiacum* and *Eurohippus messelensis*. The latter includes a pregnant mare with fetus and a specimen in which leaves and grape seeds are visible in the gut content. Skeletal material of *Hippotherium primigenium* from the Höwenegg site near Lake Constance is also on display.

Address: Hessisches Landesmuseum
Friedensplatz 1
D-64283 Darmstadt, Germany
Phone: +49 (0) 6151 165 703 or (0) 6151 165 732
E-mail: info@hlmd.de
Internet: www.hlmd.de

Dortmund, Germany

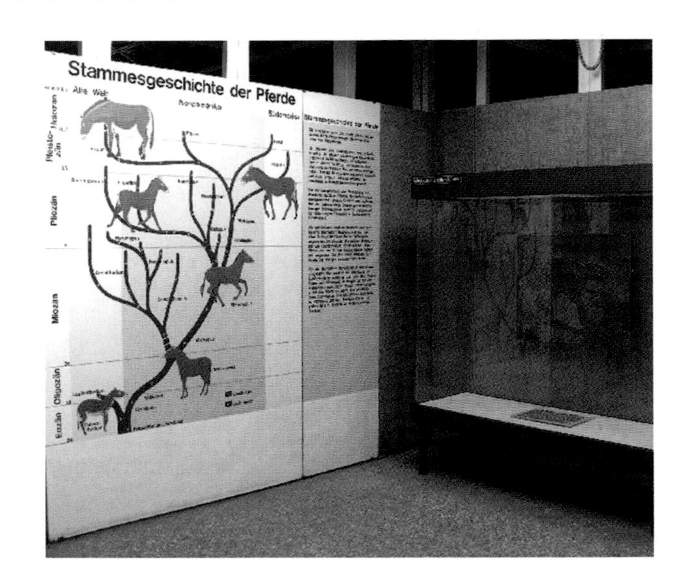

Beside a phylogenetic tree, the Museum für Naturkunde (Museum of Natural History) displays a complete, articulated skeleton of a male foal less than two months old with gut contents. It was discovered while the museum was excavating at Grube Messel in 1978. A drawing of the specimen (by Jens Franzen) is shown on signposts along highway A5 north of Darmstadt.

Address: Museum für Naturkunde
Münsterstrasse 271
D-44145 Dortmund, Germany
Phone: + 49 (0) 231 5024856
E-mail: Naturkundemuseum@stadtdo.de
Internet: www.museendortmund.de

Frankfurt am Main, Germany

The Senckenberg Museum in Frankfurt am Main is one of the most important natural history museums in Europe. The Senckenberg Research Institute has excavated at Grube Messel every year since 1975. Among the most exciting discoveries are 16 skeletons of the dawn horses *Propalaeotherium hassiacum* and *Eurohippus messelensis,* some of which are currently on exhibit. The first skeleton with gut contents (the "stamp horse") and a magnificently preserved pregnant *Eurohippus messelensis* mare with a complete fetus are among the exhibits.

Address: Forschungsinstitut und Naturmuseum
Senckenberg
Senckenberganlage 25
D-60325 Frankfurt, Germany
Phone: +49 (0) 69 7542 1357
E-mail: info@senckenberg.de
Internet: www.senckenberg.de

Halle (Saale), Germany

The somewhat antiquated but charming exhibit at the Geiseltalmuseum presents discoveries from the classic site of the Geiseltal (Geisel Valley). There are several skulls, jaws, and partial skeletons of seven different species of Middle Eocene horses that come from different levels of the former opencast lignite mine: *Hallensia matthesi, Eurohippus messelensis, E. parvulus, Propalaeotherium hassiacum, P. isselanum, P. voigti,* and *Lophiotherium sondaari*. Highlighted is the heraldic animal of the museum, a complete and articulated skeleton of *Propalaeotherium isselanum*. There is also a reconstructed skeleton of this species.

Address: Geiseltalmuseum
Domstraße 5
D-06108 Halle (Saale), Germany
Phone: + 49 (0) 345 552 6135
Internet: www.geiseltalmuseum.de

Hannover, Germany

The Niedersächsisches Landesmuseum (Country Museum of Lower Saxony) includes a beautiful diorama depicting a herd of six dawn horses of *Eurohippus messelensis*—including a foal—in different locomotory phases on the shore of Eocene Lake Messel.

Address: Niedersächsisches
Landesmuseum Hannover
Willy-Brandt-Allee 5
D-30169 Hannover, Germany
Phone: + 49 (0) 511 9807 600
Internet: www.nlmh.de

Karlsruhe, Germany

The Staatliches Museum für Naturkunde (Museum of Natural History) presents the evolution of the horses with a series of mounted skeletons (casts) of *Hyracotherium, Mesohippus, Merychippus,* and *Pliohippus,* and a skeleton of the zebra *Equus grevyi.* There are also skeletons of *Propalaeotherium hassiacum* from Messel and *Mesohippus bairdi* from the White River Formation of Wyoming. Unique to the museum is the skeleton of *Hippotherium primigenium* from the Höwenegg locality, which is shown next to a body reconstruction (dermoplastic) of a Przewalski's horse.

Address: Staatliches Museum für Naturkunde
Erbprinzenstraße 13
D-76133 Karlsruhe, Germany
Phone: + 49 (0) 721 175 2111 (entrance)
E-mail: museum@naturkundeka-bw.de
Internet: www.smnk.de

Mainz, Germany

The Naturhistorisches Museum (Museum of Natural History) in Mainz displays the body reconstructions (dermoplastics) of two adult quaggas and a foal quagga, a Somalian wild ass, and a Przewalski's horse. Other highlights are more-or-less undeformed skulls and a skeleton of *Eurohippus parvulus* from the Middle Eocene Eckfeld locality, where the museum has dug since 1987. Unique to the museum is the reconstruction of a Middle Pleistocene *Equus mosbachensis* from the Mosbach gravels and sands of the early Rhine River in the Wiesbaden vicinity.

Address: Naturhistorisches Museum Mainz
Phone: + 49 (0) 6131 122 646
E-mail: naturhistorisches.museum@stadt.mainz.de
Internet: www.uni-mainz.de/

Münster (Westphalia), Germany

Exhibitions of the horse in Germany at the Westfälisches Pferdemuseum (Westphalia Museum of Horses) at Münster focus on recent horses, particularly on equestrian sports and the successful horses and riders from Westphalia ("Westfälischer Olymp"). The presentation is introduced by a special exhibit on the evolution of horses, including a skeleton of *Eurohippus messelensis* from Grube Messel and a skeleton cast of *Palaeotherium magnum* from the Late Eocene plaster of Mormoiron (Vaucluse, southern France). The exhibit also includes intriguing visions of how horses might look 50 million years in the future.

Address: Westfälisches Pferdemuseum Münster
Sentruper Straße 311
D-48161 Münster, Germany
Phone: + 49 (0) 251 484 27 0
E-mail: info@hippomaxx-muenster.de
Internet: www.hippomaxx-muenster.de

New York, New York

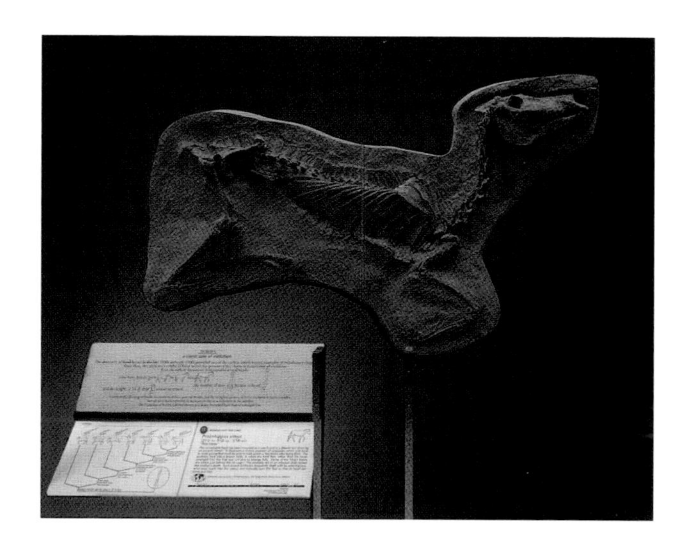

The American Museum of Natural History is one of the classic sites with exhibits on the evolution of horses. It houses the largest collection of fossil horse skeletons in the world. Here, scientists such as Walter Granger, William D. Matthew, and George Gaylord Simpson—to name only a few—worked. The exhibit shows the linear development of skeletons from *Hyracotherium* (*Eohippus*) to modern *Equus*, with specimens gradually getting larger, with fewer toes and higher teeth (anagenesis). Another arrangement follows cladistic systematics, focusing on the splitting of species (cladogenesis), which resulted in the systematic arrangement of organisms by shared, specialized features (synapomorphic characters). Together the arrangements make it clear that evolution is a complex, branching story. For example some later horses, such as *Calippus*, are smaller than earlier ones, while others, such as *Neohipparion*, still had three toes.

Address: The American Museum of Natural History
Central Park West at 79th Street
New York, N.Y. 10024-5192
Phone: (212) 769-5100
Internet: www.amnh.org

Stuttgart, Germany

The Staatliches Museum für Naturkunde (Museum am Löwentor) presents two skeletons—a magnificent foal and a subadult individual—of *Eurohippus messelensis* from Grube Messel in the context of horse evolution. The museum also has the only mounted skeleton (cast) of *Hippotherium primigenium* from the Höwenegg locality and a skeleton of the palaeothere *Plagiolophus minor* assembled from isolated bones from the Frohnstetten locality on the Suebian Alb.

Address: Staatliches Museum für Naturkunde
(Museum am Löwentor)
Rosenstein 1
D-70191 Stuttgart, Germany
Phone: + 49 (0) 711 8936 0
E-mail: SMNS@naturkundemuseum-bw.de
Internet: www.naturkundemuseum-bw.de

Verden (Aller), Germany

The Deutsches Pferdemuseum (German Museum of Horses) is located in the stables of a former cavalry garrison. The exposition provides an overview of horse evolution with the help of a series of mounted skeletons (casts) ranging from *Hyracotherium* to *Mio-* and *Merychippus* to the genus *Equus*. A cast of *Eurohippus messelensis* from Grube Messel and a special exhibition on the evolutionary development of horses' feet are also on display.

Address: Deutsches Pferdemuseum
Holzmarkt 9
D-27283 Verden, Germany
Phone: + 49 (0) 4231 807 140
E-mail Pferdemuseum@t-online.de
Internet: www.dpm-verden.de

FURTHER READING

General Overview

Azzaroli A (1985) An early history of horsemanship. E. J. Brill/Dr. W. Backhuys, Leiden

Budiansky S (1997) The nature of horses. Exploring equine evolution, intelligence, and behavior. Free Press, New York

Franzen JL (2002) Die Evolution der Pferde. In Marx C, Sternschulte A (eds.) So frei, so stark. Westfalens wilde Pferde. Klartext Verlag, Essen: 59–68

Franzen JL (2007) Die Urpferde der Morgenröte. Ursprung und Evolution der Pferde. Spektrum Akademischer Verlag (Elsevier), Heidelberg

MacFadden BJ (1992) Fossil horses. Systematics, paleobiology, and evolution of the family Equidae. Cambridge University Press, Cambridge

Prothero DR, Schoch RM (1989) The evolution of peerissodactyls. Clarendon Press, Oxford University Press, New York, Oxford

Prothero DR, Schoch RM (2002) Horns, tusks, and flippers. The evolution of hoofed mammals. Johns Hopkins University Press, Baltimore

Schaal S (ed.) (2005) UNESCO-Welterbe Fossilienlagerstätte Grube Messel. *Vernissage, Reihe: Unesco-Welterbe* 13 (21–05): 1–67

Simpson GG (1977) Pferde. Die Geschichte der Pferdefamilie in der heutigen Zeit und in sechzig Millionen Jahren ihrer Entwicklung. Verlag Paul Parey, Berlin und Hamburg. (The original English edition, *Horses,* was published in 1951 by Oxford University Press)

Thenius E, Hofer H (1960) Stammesgeschichte der Säugetiere. Springer-Verlag, Berlin, Göttingen, Heidelberg

Thenius E (1969) Stammesgeschichte der Säugetiere (einschließlich der Hominiden). Handbuch der Zoologie 8.2. de Gruyter, Berlin

1. Prologue

Behnke C, Eikamp H, Zollweg M (1986) Die Grube Messel. Goldschneck-Verlag, Korb

Franzen JL (1975) Müll in paläontologische Fundgrube? *Umschau in Wissenschaft und Technik* 75 (3): 68

Franzen JL (1976) Senckenbergs Grabungskampagne 1975 in Messel: Erste Ergebnisse und Ausblick. *Natur und Museum* 106 (7): 217–223

Franzen JL (1977) Urpferdchen und Krokodile: Messel vor 50 Millionen Jahren. *Kleine Senckenberg-Reihe* 7: 1–6 (New editions in 1979 and 1982)

Franzen JL (1978) Messel-Fossilien auf Briefmarken. *Natur und Museum* 108 (9): 281–283

Franzen JL (1988) Die Grube Messel und der Müll. *Natur und Museum* 118 (5): 133–142

Klausewitz W (2000) Die frühen Bemühungen Senckenbergs um die Grube Messel. Versuch einer historischen Dokumentation. *Natur und Museum* 130 (10): 333–347

2. Introduction

Chauvet JM, Brunel Deschamps E, Hillaire C (2001) Grotte Chauvet bei Vallon-Pont-d'Arc. Altsteinzeitliche Höhlenkunst im Tal der Ardèche. Jan Thorbecke Verlag, Stuttgart

Clottes J (ed.) (2001) La Grotte Chauvet. L'Art des Origines. Editions du Seuil, Paris

Kidd J (1985) Horse breeds & breeding. Salamander Books, London

Lorblanchet M (2000) Höhlenmalerei. Ein Handbuch. 2. A., Jan Thorbecke Verlag, Stuttgart

McKenna MC, Bell SK (1997) Classification of mammals above the species level. Columbia University Press, New York, Chichester

Outram AK, Stear NA, Bendrey R, Olsen S, Kasparov A, Zaibert V, Thorpe N, and Evershed RP (2009): The earliest horse harnessing and milking. *Science* 323: 1332–1335.

Römisch-Germanisches Zentralmuseum, Forschungsinstitut für Vor- und Frühgeschichte (ed.) (1999) The role of early humans in the accumulation of European lower and middle palaeolithic bone assemblages. Ergebnisse eines Kolloquiums. *Monographien Römisch-Germanisches Zentralmuseum* 42: 1–396

Steinbach G (1979) Die Pferde. Deutscher Bücherbund, Stuttgart, Hamburg, München

Thieme H (1997) Lower Palaeolithic hunting spears from Germany. *Nature* 385: 807–810

Thieme H (1999) Altpaläolithische Holzgeräte aus Schöningen, Lkr. Helmstedt. Bedeutsame Funde zur Kulturentwicklung des frühen Menschen. *Germania* 77: 451–487

3. The Depths of Time

Franzen JL (1988) Geologisch-paläontologische Methoden der Chronologie. In Knussmann R (ed.) Anthropologie. Handbuch der vergleichenden Biologie des Menschen, 1: Wesen und Methoden der Anthropologie. Gustav Fischer, Stuttgart, New York: 642–655

Franzen JL, Weidmann M, Berger J-P (1991) Sur l'âge des Couches à Cérithes ou Couches des Diablerets de l´Eocène alpin. *Eclogae geologicae Helvetiae* 84 (3): 893–919

Geyh MA (1988) Chronologische Methoden der Physik und Chemie. In Knussmann R (ed.) Anthropologie. Handbuch der vergleichenden Biologie des Menschen 1: Wesen und Methoden der Anthropologie. Gustav Fischer, Stuttgart, New York: 670–683

Geyh MA, Schleicher H (1990) Absolute age determination. Physical and chemical dating methods and their application. Springer-Verlag, Berlin, Heidelberg

Kappelman J, Sen S, Fortelius M, Duncan A, Alpagut B, Crabaugh J, Gentry A, Lunkka JP, McDowell F, Solounias N, Viranta S, Werdelin L (1996) Chronology and biostratigraphy of the Miocene Sinap Formation of central Turkey. In Bernor RL, Fahlbusch V, Mittmann H-W (eds.) The evolution of western Eurasian Neogene mammal faunas. Columbia University Press, New York: 78–95

Koch PL, Clyde WC, Hepple RP, Fogel ML, Wing SL, Zachos JC (2003) Carbon oxygen isotope records from paleosols spanning the Paleocene-Eocene boundary, Bighorn Basin, Wyoming. In Wing SL, Gingerich PD, Schmitz B, Thomas E (eds.) Causes and consequences of globally warm climates in the early Paleogene. *Special Paper of the Geological Society of America* 369: 49–64

Norman JE, Ashley MV (2000) Phylogenetics and tests of the molecular clock. *Journal of Molecular Evolution* 50: 11–21

Press F, Siever R (1995) Allgemeine Geologie. Spektrum Akademischer Verlag, Heidelberg, Berlin, Oxford. (The original U.S. edition was published in 1994 by W. H. Freeman)

4. Europe in the Eocene

Gingerich PD, Ul Haq M, Zamout IS, Khan ICH, Malkani MS (2001) Origin of whales from early artiodactyls: hands and feet of Eocene Protocetidae from Pakistan. *Science* 293: 2239–2242

Harms F-J (2002) Steine erzählen Geschichte(n): Ursache für die Entstehung des Messel-Sees gefunden. *Natur und Museum* 132 (1): 1–4

Lutz H, Neuffer FO, Harms F-J, Schaal S, Micklich N, Gruber G, Schweigert G, Lorenz V (2000) Tertiäre Maare als Fossillagerstätten: Eckfeld, Messel, Randeck, Höwenegg, Öhningen. Tertiary maars as fossil deposits: Eckfeld, Messel, Randeck, Höwenegg, Öhningen. *Mainzer Naturwissenschaftliches Archiv* Suppl. 24: 125–160

Meischner D (2002) Europäische Fossillagerstätten. Springer, Berlin, Heidelberg

Mertz DF, Renne PR (2005) A numerical age for the Messel fossil deposit (UNESCO World Heritage Site) derived from 40Ar/39Ar dating on a basaltic rock fragment. *Courier Forschungsinstitut Senckenberg* 255: 67–75

Mertz DF, Swisher CC, Franzen JL, Neuffer O, Lutz H (2000) Numerical dating of the Eckfeld maar fossil site, Eifel, Germany: a calibration mark for the Eocene time scale. *Naturwissenschaften* 87: 270–274

Thewissen JGM, Williams EM, Roe LJ, Hussain ST (2001) Skeletons of terrestrial cetaceans and the relationship of whales to artiodactyls. *Nature* 413: 277–281

5. The Dawn Horses of the Morning Cloud

Franzen JL (1985) Exceptional preservation of Eocene vertebrates in the lake deposit of Grube Messel (West Germany). *Philosophical Transactions of the Royal Society London* (B) 311: 181–186

Franzen JL (1990) *Hallensia* (Mammalia, Perissodactyla) aus Messel und dem Pariser Becken sowie Nachträge aus dem Geiseltal. *Bulletin de l'Institut Royal des Sciences Naturelles de Belgique* 60: 175–201

Franzen JL (2006) *Eurohippus* n.g., a new genus of horses from the Middle to Late Eocene of Europe. *Senckenbergiana lethaea* 86 (1): 97–102

Franzen JL (2007) Eozäne Equoidea (Mammalia, Perissodactyla) aus der Grube Messel bei Darmstadt (Deutschland). Funde der Jahre 1969–2000. Schweizerische Paläontologische Abhandlungen 127: 1–245, 41 Abb., Taf. 1–61, 57 Tab.; Kommission der Schweizerischen Paläontologischen Abhandlungen, Basel

Franzen JL, Frey E (1990) *Europolemur* completed. *Kaupia, Darmstädter Beiträge zur Naturgeschichte* 3: 113–130

Habersetzer J, Schaal S (eds.) (2004) Current geological and paleontological research in the Messel Formation. *Courier Forschungsinstitut Senckenberg* 252: 1–245

Harms F-J, Buness H, Felder M, Nix T (2004) Neue Vorstellungen zur Entstehung der Welterbestätte Grube Messel in aktuellen geologischen und seismischen Schnitten. *Bericht Naturwissenschaftlicher Verein Darmstadt* (NF) 27: 71–86

Harms F-J, Nix T, Felder M (2003) Neue Darstellungen zur Geologie des Ölschiefer-Vorkommens Grube Messel. *Natur und Museum* 133 (5): 140–148

Haubold H (1995) Wirbeltiergrabung und -forschung im Geiseltaleozän. *Hallesches Jahrbuch für Geowissenschaften, Reihe B: Geologie, Paläontologie, Mineralogie* 17: 1–18

Haupt O (1925) Die Palaeohippiden der eozänen Süßwasserablagerungen von Messel bei Darmstadt. *Abhandlungen der Hessischen Geologischen Landesanstalt* 6 (4): 1–159

Heil R, Koenigswald W von, Lippmann HG, Graner D, Heunisch C (1987) Fossilien der Messel-Formation. Hessisches Landesmuseum, Darmstadt

Hellmund M, Koehn C (2000) Skelettrekonstruktion von *Propalaeotherium hassiacum* (Equidae, Perissodactyla, Mammalia) basierend auf Funden aus dem eozänen Geiseltal (Sachsen-Anhalt, Deutschland). *Hallesches Jahrbuch für Geowissenschaften*. Series B, Suppl. 12: 1–55

Koenigswald W von (1987) Die Fauna des Ölschiefers von Messel. In Heil R, Koenigswald W von, Lippmann HG, Graner D, Heunisch C (eds.) Fossilien der Messel Formation. Hessisches Landesmuseum, Darmstadt: 71–142

Koenigswald W von, Braun A, Pfeiffer T (2004) Cyanobacteria and seasonal death. A new taphonomic model for the Eocene Messel lake. *Paläontologische Zeitschrift* 78 (2): 417–424

Koenigswald W von, Schaarschmidt F (1983) Ein Urpferd aus Messel, das Weinbeeren fraß. *Natur und Museum* 113 (3): 79–84

Koenigswald W von, Storch G (eds.) (1998) Messel. Ein Pompeji der Paläontologie. Jan Thorbecke Verlag, Sigmaringen

Krumbiegel G, Rüffle L, Haubold H (1983) Das eozäne Geiseltal, ein mitteleuropäisches Braunkohlenvorkommen und seine Tier- und Pflanzenwelt. A. Ziemsen, Wittenberg Lutherstadt

Matthes HW (1977) Die Equiden aus dem Eozän des Geiseltales. 1. Die Zähne. In Matthes HW, Thaler B (eds.) Eozäne Wirbeltiere des Geiseltales. *Wissenschaftliche Beiträge, Martin-Luther-Universität Halle-Wittenberg* (1977) 2 (P5): 5–39, and plates 1–176

Morlo M, Schaal S, Mayr G, Seiffert C. (2004) An annotated taxonomic list of the Middle Eocene (MP 11) vertebrata of Messel. *Courier Forschungs-Institut Senckenberg* 252: 95–108

Neuffer FO, Gruber G, Lutz H, Frankenhäuser H (1996) Das Eckfelder Maar. Zeuge tropischen Lebens in der Eifel. Ein Bildband zu den bis 1995 bekannt gewordenen Pflanzen- und Tier-Arten mit Erläuterungen zu Geologie und Paläogeographie. Naturhistorisches Museum Mainz, Landessammlung für Naturkunde Rheinland-Pfalz Mainz

Richter G (1988) Versteinerte Magen/Darm-Inhalte, ihre Analyse und Deutung. In Schaal S, Ziegler W (eds.) Messel: Ein Schaufenster in die Geschichte der Erde und des Lebens. Verlag Waldemar Kramer, Frankfurt am Main: 285–289

Savage DE, Russell DE, Louis P (1965) European Eocene Equidae (Perissodactyla). *University of California Publications in Geological Sciences* 56: 1–94

Schaal S, Schneider U (1995) Chronik der Grube Messel. Verlag Kempkes, Gladenbach

Schaal S, Ziegler W (eds.) (1992) Messel: An insight into the history of life and of the Earth. Clarendon Press, Oxford

Wolf HW (1988) Schätze im Schiefer. Faszinierende Fossilien aus der Grube Messel. Westermann, Braunschweig

Wuttke M (1983) "Weichteil-Erhaltung" durch lithifizierte Mikroorganismen bei mittel-eozänen Vertebraten aus den Ölschiefern der "Grube Messel" bei Darmstadt. *Senckenbergiana lethaea* 64 (5–6): 509–527

6. Constructions and Functions

Camp CL, Smith N (1942) Phylogeny and functions of the digital ligaments of the horse. *Memoirs University of California* 13: 69–122

Darwin C (1860) On the origin of species by means of natural selection. 2nd ed., Oxford University Press, Oxford (reprint)

Darwin C (1962) Reise eines Naturforschers um die Welt. Steingrüben Verlag, Stuttgart

Ellenberger W, Dittrich H, Baum H (1956) An atlas of animal anatomy for artists. Dover Publications, New York

Frankfurter Evolutionstheorie: Available on the Internet at www.evolutionswissenschaften.de/forsch_bio_FET.html

Gutmann WF (1989) Evolution hydraulischer Konstruktionen. Organismische Wandlung statt altdarwinistischer Anpassung. Kramer, Frankfurt

Gutmann WF, Bonik K (1981) Kritische Evolutionstheorie. Ein Beitrag zu Überwindung altdarwinistischer Dogmen. Gerstenberg, Hildesheim

Hermanson JW, MacFadden BJ (1992) Evolutionary and functional morphology of the shoulder joint and stay-apparatus in fossil and extant horses (Equidae). *Journal of Vertebrate Paleontology* 12 (3): 377–386

Hertler C (2001) Morphologische Methoden in der Evolutionsforschung. *Studien zur Theorie der Biologie* 5: 1–364

Hildebrand M (1987) The mechanics of horse legs. *American Scientist* 75: 594–601

Hulbert RC (1993) Late Miocene *Nannippus* (Mammalia, Perissodactyla) from Florida, with a description of the smallest hipparionine horse. *Journal of Vertebrate Paleontology* 13: 350–366

MacFadden BJ, Cerling TE (1994) Fossil horses, carbon isotopes, and global change. *Trends in Ecological Evolution* 9: 481–485

Matthew WD (1926) The evolution of the horse. A record and its interpretation. *Quarterly Review of Biology* 1 (2): 139–185

Owen-Smith R (1988) Megaherbivores: the influence of very large body size on ecology. Cambridge University Press, Cambridge

Peters DS, Gutmann WF (1971) Über die Lesrichtung von Merkmals- und Konstruktionsreihen. *Zeitschrift für zoologische Systematik und Evolutionsforschung* 9: 237–263

Pfretzschner HU (1993) Enamel microstructure in the phylogeny of the Equidae. *Journal of Vertebrate Paleontology* 13 (3): 342–349

Piperno R, Sues H-D (2005) Dinosaurs dined on grass. *Science* 310 (5751): 1126–1128

Prasad V, Strömberg CAE, Alimohammadian H, Sahni A (2005) Dinosaur coprolites and the early evolution of grasses and grazers. *Science* 310 (5751): 1177–1180

Renders E (1984) The gait of *Hipparion* sp. from fossil footprints in Laetoli, Tanzania. *Nature* 308: 179–181

Solounias N, Semprebon G (2002) Advances in the reconstruction of ungulate ecomorphology with application to early fossil equids. *American Museum Novitates* 3366: 1–49

Sondaar PY (1968) The osteology of the manus of fossil and recent Equidae with special reference to phylogeny and function. *Verhandelingen der Koninklijke Nederlandse Akademie van Wetenschappen, Natuurkunde* 25 (1): 1–76

Strömberg CAE (2002) The origin and spread of grass-dominated ecosystems in the late Tertiary of North America: preliminary results concerning the evolution of hypsodonty. *Palaeogeography, Palaeoclimatology, Palaeoecology* 177: 59–75

Strömberg CAE (2004a) Using phytolith assemblages to reconstruct the origin and spread of grass-dominated habitats in the Great Plains of North America during the late Eocene to early Miocene. *Palaeogeography, Palaeoclimatology, Palaeoecology* 207: 239–275

Strömberg CAE (2004b) Decoupled taxonomic radiation and ecological expansion of open-habitat grasses in the Cenozoic of North America. *PNAS* 102 (34): 11980–11984

Tobien H (1952) Über die Funktion der Seitenzehen tridactyler Equiden. *Neues Jahrbuch für Geologie und Paläontologie, Abhandlungen* 96: 137–172

Wallace AR (1855) On the law which has regulated the introduction of new species. *Annals and Magazine of Natural History* (2) 16: 184–196

Wang Y, Cerling TE, and MacFadden BJ (1994) Fossil horses and carbon isotopes: new evidence for Cenozoic dietary, habitat, and ecosystem changes in North America. *Palaeogeography, Palaeoclimatology, Palaeoecology* 107: 269–280

7. Discovering Horse Evolution

Bernor RL, Tobien H, Hayek L-A, Mittmann H-W (1997) *Hippotherium primigenium* (Equidae, Mammalia) from the late Miocene of Höwenegg (Hegau, Germany). *Andrias* 10: 1–230

Brown LS (ed.) (1957) Muybridge, Eadweard (1899) Animals in motion. Dover Publications, New York

Edinger T (1948) Evolution of the horse brain. *The Geological Society of America, Memoir* 25: I–X, 1–177

Franzen JL (1984) Die Stammesgeschichte der Pferde in ihrer wissenschaftshistorischen Entwicklung. *Natur und Museum* 114 (6): 149–162

Froehlich DJ (2002) Quo vadis *eohippus*? The systematics and taxonomy of the early Eocene equids (Perissodactyla). *Zoological Journal of the Linnean Society* 134: 141–256

Gingerich PD (1980) History of early cenozoic vertebrate paleontology in the Bighorn Basin. In Gingerich PD (ed.) Early Cenozoic Paleontology and Stratigraphy of the Bighorn Basin, Wyoming. *Papers on Paleontology* 24: I–VI, 1–146

Huxley TH (1876) Lectures on evolution. *Collected Essays* vol. 4

Kaiser T (2003) The dietary regimes of two contemporaneous populations of *Hippotherium primigenium* (Perissodactyla, Equidae) from the Vallesian (Upper Miocene) of Southern Germany. *Palaeogeography, Palaeoclimatology, Palaeoecology* 198: 381–402

Kohring R (1997) Senckenbergische Forscher: Tilly Edinger (1897–1967). *Natur und Museum* 127 (11): 391–410

Kowalevsky WA (1873a) Sur *l'Anchitherium aurelianense* Cuv. et sur l'histoire paléontologique des chevaux. *Mémoires de l'Académie Impériale des Sciences, St. Petersbourg* (7) 20 (5): 1–73

Kowalevsky WA (1873b) On the osteology of the Hyopotamidae. *Philosophical Transactions of the Royal Society of London* 163: 19–94

Kowalevsky WA (1876) Monographie der Gattung *Anthracotherium* Cuv. und Versuch einer natürlichen Classifikation der fossilen Hufthiere. *Palaeontographica* (NF) 22: 133–347

Muybridge E (1899) Animals in motion. Chapman and Hall, London

Radinsky L (1976) Oldest horse brains: more advanced than previously realized. *Science* 194: 626–627

Schwartze S, Seidel R, Schilling N, Fischer MS, Haas A (2001) Bringing fossils back to life: the locomotion of the Messel horse, *Propalaeotherium parvulum*. *Journal of Morphology* 248 (3), *Special Issue: Sixth International Congress of Vertebrate Morphology—Jena, Germany, July 2–26, 2001*: 282 (abstract)

Sternberg CH (1990) The life of a fossil hunter. Indiana University Press, Bloomington and Indianapolis

Strelnikov I, Hecker R (1968) Wladimir Kowalevsky's sources of ideas and their importance for his work and for Russian evolutionary paleontology. *Lethaia* 1: 219–229

8. Evolution and Expansion of the Horses

Abusch-Siewert S (1983) Gebißmorphologische Untersuchungen an eurasiatischen Anchitherien (Equidae, Mammalia) unter besonderer Berücksichtigung der Fundstelle Sandelzhausen. *Courier Forschungsinstitut Senckenberg* 62: 1–401

Badiola A, Checa L, Cuesta MA, Quer R, Hooker JJ, and Astibia H. (2009) The role of New Iberian finds in understanding European Eocene mammalian palaeobiogeography. *Geologica Acta* 7 (1–2)

Bains S, Norris RD, Corfield RM, Bowen GJ, Gingerich PD, Koch PL (2003) Marine-terrestrial linkages at the Paleocene-Eocene boundary. In Wing SL, Gingerich PD, Schmitz B, Thomas E (eds.) Causes and consequences of globally warm climates in the early Paleogene. *Special Paper of the Geological Society of America* 369: 1–9

Bernor RL (1999) Family Equidae. In Rössner GE, Heissig K (eds.) Land mammals of Europe. Verlag Dr. Friedrich Pfeil, München: 193–202

Bernor RL, Koufos GD, Woodburne MO, Fortelius M (1996) The evolutionary history and biochronology of European and southwest Asian late Miocene and Pliocene Hipparionine horses. In Bernor RL, Fahlbusch V, Mittmann H-W (eds.) The evolution of western Eurasian Neogene mammal faunas. Columbia University Press, New York: 307–338

Checa L, Colombo F (2004) A new early Eocene palaeothere (Mammalia, Perissodactyla) from Northeastern Spain. *Journal of Vertebrate Paleontology* 24 (2): 510–515

Dashzeveg D (1979a) On an archaic representative of the equoids (Mammalia, Perissodactyla) from the Eocene of central Asia. *Transactions of the Joint Soviet-Mongolian Paleontological Expedition* 8: 10–22

Dashzeveg D (1979b) Der Fund eines Hyracotheriums in der Mongolei. *Paläontologisches Journal* 3: 108–113 (in Russian)

Deng Tao, Xue Xiang-Xu (1999) Chinese fossil horses of *Equus* and their environment. China Ocean Press (in Chinese with English summary)

Eisenmann V (1998) Quaternary horses: possible candidates to domestication. *Proceedings of the XIII International Congress of Prehistoric and Protohistoric Sciences, Forli, Italia, 8–14 September 1996* 6, (1), Workshop 3. The horse: its domestication, diffusion and role in past communities: 27–36

Gingerich PD (1989) New earliest Wasatchian mammalian fauna from the Eocene of northwestern Wyoming: composition and diversity in a rarely sampled high-floodplain assemblage. *University of Michigan Papers on Paleontology* 28: 1–97

Gingerich PD (2003) Mammalian responses to climate change at the Paleocene-Eocene boundary: Polecat Bench record in the northern Bighorn Basin, Wyoming. In Wing SL, Gingerich PD, Schmitz B, Thomas E (eds.) Causes and consequences of globally warm climates in the early Paleogene. *Special Paper of the Geological Society of America* 369: 463–478

Hooker JJ (1994) The beginning of the equoid radiation. *Zoological Journal of the Linnean Society* 112: 29–63

Miao D-S (1994) Early Tertiary fossil mammals from the Shiao Basin, Panxian County, Guizhou Province. *Acta Palaeontologica Sinica* 21 (5): 526–536 (in Chinese with English summary)

Osborn HF (1919) Equidae of the Oligocene, Miocene, and Pliocene of Northamerica. Iconographic type revision. *Memoirs of the American Museum of Natural History* (NS) 2: 1–326

Russell DE, Zhai R (1987) The Paleogene of Asia: mammals and stratigraphy. Éditions du Muséum National d'Histoire Naturelle, Paris

Storch G (1986) Die Säuger von Messel: Wurzeln auf vielen Kontinenten. Spektrum der Wissenschaft 6 (1986): 48–65

Ting S, Bowen GJ, Koch PL, Clyde WC, Wang Y, McKenna MC (2003) Biostratigraphic, chemostratigraphic, and magnetostratigraphic study across the Paleocene-Eocene boundary in the Hengyang Basin, Hunan, China. In Wing SL, Gingerich PD, Schmitz B, Thomas E (eds.) Causes and consequences of globally warm climates in the early Paleogene. *Special Paper of the Geological Society of America* 369: 521–535

Zdansky O (1930) Die alttertiären Säugetiere Chinas nebst stratigraphischen Bemerkungen. *Palaeontologia Sinica* (C) 6 (2): 1–87

9. Pseudo Horses and Relatives of Horses

Coombs MC (1989) Interrelationships and diversity in the Chalicotheriidae. In Prothero DR, Schoch RM (eds.) The evolution of perissodactyls. Clarendon Press, Oxford University Press, New York, Oxford: 438–457

Franzen JL (1995) Die Equoidea des europäischen Mitteleozäns. *Hallesches Jahrbuch für Geowissenschaften. Reihe B: Geologie, Paläontologie, Mineralogie* 17: 31–45

Froehlich DJ (1999) Phylogenetic systematics of basal perissodactyls. *Journal of Vertebrate Paleontology* 19 (1): 140–159

Guerin C (1980) Les rhinocéros (Mammalia, Perissodactyla) du Miocène Terminal au Pleistocène Supérieur en Europe Occidentale. Comparaison avec les espèces actuelles. *Documents des Laboratoires de Géologie Lyon* 79 (1–3) : 1–1185

Heissig K (1999a) Family Tapiridae. In Rössner GE, Heissig K (eds.) Land mammals of Europe. Verlag Dr. Friedrich Pfeil, München: 171–174

Heissig K (1999b) Family Rhinocerotidae. In Rössner GE, Heissig K (eds.) Land mammals of Europe. Verlag Dr. Friedrich Pfeil, München: 175–188

Heissig, K. (1999c) Family Chalicotheriidae. In Rössner GE, Heissig K (eds.) Land mammals of Europe. Verlag Dr. Friedrich Pfeil, München: 189–192

Holbrook LT (2001) Comparative osteology of early Tertiary tapiromorphs (Mammalia, Perissodactyla). *Zoological Journal of the Linnean Society* 132: 1–54

Hooker JJ (1984) A primitive ceratomorph (Perissodactyla, Mammalia) from the early Tertiary of Europe. *Zoological Journal of the Linnean Society* 82: 229–244

Janis CM (1976) The evolutionary strategy of the Equidae and the origin of rumen and cecal digestion. *Evolution* 30: 757–774

Lucas SG, Holbrook LT (2004) The skull of the Eocene perissodactyl *Lambdotherium* and its phylogenetic significance. In Lucas SG, Zeigler KE, Kondrashov PE (eds.) Paleogene mammals. *Science Bulletin* 26: 81–88

Osborn HF (1929) The Titanotheres of ancient Wyoming, Dakota, and Nebraska. *United States Geological Survey Monograph* 55: 1–953

Radinsky LB (1965) Evolution of the Tapiroid skeleton from *Heptodon* to *Tapirus*. *Bulletin of the Museum of Comparative Zoology* 134 (3): 69–106

Radinsky LB (1967) A review of the Rhinocerotoid family Hyracodontidae (Perissodactyla). *Bulletin of the American Museum of Natural History* 136 (1): 1–46

Radinsky LB (1969) The early evolution of the Perissodactyla. *Evolution* 23 (2): 308–328

Zapfe H (1979) *Chalicotherium grande* (Blainv.) aus der miozänen Spaltenfüllung von Neudorf an der March (Děvinská Nová Ves), Tschechoslowakei. Verlag Ferdinand Berger & Söhne, Wien-Horn

10. The Ice Age and the Roots of Modern Horses

Berner U, Streif H (2000) Klimafakten. Der Rückblick: Ein Schlüssel für die Zukunft. 2. verbesserte A., Bundesanstalt für Geowissenschaften und Rohstoffe, Hannover

Clutton-Brock J (1992) Horse power. A history of the horse and donkey in human societies. Natural History Museum Publications, London.

Eisenmann V (1980) Les chevaux (*Equus* sensu lato) fossiles et actuels: crânes et dents jugales supérieures. *Cahiers de Paléontologie CNRS*: 1–186

Eisenmann V (1998) Quaternary horses: possible candidates to domestication. *Proceedings of the XIII International Congress of Prehistoric and Protohistoric Sciences, Forli, Italia, 8–14 September 1996* 6 (1), Workshop 3. The horse: its domestication, diffusion, and role in past communities: 27–36

Eisenmann V, Baylac M (2000) Extant and fossil *Equus* (Mammalia, Perissodactyla) skulls: a morphometric definition of the subgenus *Equus*. *Zoologica Scripta* 29 (2): 89–100

Eisenmann V, Kuznetsova T (2004) Early Pleistocene equids (Mammalia, Perissodactyla) of Nalaikha, Mongolia, and the emergence of modern *Equus* Linnaeus, 1758. *Geodiversitas* 26 (3): 535–561

Groves CP (1994) The Przewalski horse: morphology, habitat and taxonomy.

In Boyd L, Houpt KA (eds.) Przewalski's horse: The history and biology of an endangered species. State University of New York Press, Albany

MacFadden, BJ, Carranza-Castañeda O (2002) Cranium of *Dinohippus mexicanus* (Mammalia: Equidae) from the early Pliocene of Central Mexico, and the origin of *Equus*. *Bulletin of the Florida Museum of Natural History* 43: 163–185

Oakenfull EA, Lim HN, Ryder O (2000) A survey of equid mitochondrial DNA: implications for the evolution, genetic diversity and conservation of *Equus*. *Conservation Genetics* 1: 341–355

Sher AV (1992) Beringian fauna and Early Quaternary mammalian dispersal in Eurasia: ecological aspects. *Courier Forschungsinstitut Senckenberg* 153: 125–133

Spassov N, Iliev N (1997) The wild horses of Eastern Europe and the polyphyletic origin of the domestic horse. *Anthropozoologica* 25–26: 753–761

Uerpmann H-P (1995) Domestication of the horse: when, where, and why ? In Le cheval et les autres équidés: aspects de l'histoire de leur insertion dans les activités humaines. *Colloques d'histoire des connaissances zoologiques* 6: 15–29

Weinstock J, Willerslev E, Sher A, Tong W, Ho SYW, Rubinstein D, Storer J, Burns J, Martin L, Bravi C, Prieto A, Froese D, Scott E, Xulong L, Cooper A (2005) Evolution, systematics, and phylogeography of Pleistocene horses in the New World: a molecular perspective. *PLoS Biology* 3 (8): 1–7

11. Conclusion

Estravis C (1992) Estudo dos Mamíferos do Eocénico inferior de Silveirinha. (Baixo Mondego). Thesis doctoral. Faculdade de Ciéncias e Tecnologia, UNL, Lisboa: 1–254

Fischer M (1986) Die Stellung der Schliefer (Hyracoidea) im phylogenetischen System der Eutheria. Zugleich ein Beitrag zur Anpassungsgeschichte der Procaviidae. *Courier Forschungsinstitut Senckenberg* 84: 1–132

Gingerich PD (1990) African dawn for primates. *Nature* 346: 411

MacFadden BJ (2005) Fossil horses: evidence for evolution. *Science* 307: 1728–1730

Sigé B, Jaeger J-J, Sudre J, Vianey-Liaud M (1990) *Altiatlasius koulchii* n. gen. et sp., Primate omomyidé du Paléocène Supérieur du Maroc, et les origines des Euprimates. *Palaeontographica* A 214: 31–56

Exhibits on the Evolution of Horses

Deutsches Pferdemuseum (ed.) (2003) Pferde. Geschichte und Geschichten. Deutsches Pferdemuseum, Verden (Aller)

Ebers C, Ebers S, Encke D, Esch C, Hermanns U, Rasbach P, Stoffregen-Büller M, Wecker I (2004) Von Pferden und Menschen. Das Westfälische Pferdemuseum im Allwetterzoo Münster. Landwirtschaftsverlag, Münster

Pucka G (1994) Zur Entstehung eines Urpferdchendioramas im Niedersächsischen Landesmuseum Hannover. *Der Präparator* 40(4):139–142

INDEX

Page numbers followed by *f* refer to figures.

Illustration Credits

1. Prologue
1.1 © Hessisches Landesmuseum Darmstadt, photo: Wolfgang Fuhrmannek
1.2 © Author
1.3 © Forschungsinstitut und Naturmuseum Senckenberg Frankfurt, photo: Christel Schumacher
1.4 © Hessisches Landesmuseum Darmstadt, photo: Dieter Keller, Mühltal
1.5 © Author
1.6 © Author
1.7 © Author
1.8 © Author
1.9 © Author
1.10 © Forschungsinstitut und Naturmuseum Senckenberg Frankfurt, photo: Erwin Haupt
1.11 © Author
1.12 © Forschungsinstitut und Naturmuseum Senckenberg Frankfurt, photo: Sven Tränkner
1.13 © Forschungsinstitut und Naturmuseum Senckenberg Frankfurt, photo: Christel Schumacher

2. Introduction
2.1 Vera Kassühlke (design by the author)
2.2 Vera Kassühlke (after Steinbach, Die Pferde: 73)
2.3 Vera Kassühlke
2.4 © Author (after MacFadden 1992: 85, Fig. 5.3)
2.5 a: © Author
2.5 b: © Author (Used by permission of Erlaubnis des Allwetterzoo Münster)
2.5 c: © Carl Hagenbeck Archives, photo: Dr. Stephan Hering-Hagenbeck
2.5 d: © Author (Used by permission of Erlaubnis des Allwetterzoo Münster)
2.6 a and c: © Author (Used by permission of the Allwetterzoo in Münster)
2.6 b: © Prof. Dr. Rainer Willmann; Göttingen
2.6 d: Photo in the public domain, Wacholderhain, Haselünne
2.7 © Author
2.8 © 2005 Dr. Hartmut Thieme, Niedersächsisches Landesamt für Denkmalpflege, Hannover, photo: Peter Pfarr
2.9 © Centre de Préhistoire du Pech Merle, Cabrerets (France); photo: P. Cabrol
2.10 © Author
2.11 © Ethnographic Museum Stockholm, photo: Roland Reed
2.12 © Musée de la Tapisserie Bayeux; reproduced by special permission of the city of Bayeaux
2.13 © Deutsches Pferdemuseum Verden, photo: Michael Hensel
2.14 © Deutsches Pferdemuseum Verden, painting from: Franz Krüger 1841 (on loan from Ernst August); photo: Michael Hensel
2.15 © Swiss National Circus, Circus Knie; photographer: Geri Kuster
2.16 © Pferdebild-Agentur Werner Ernst, Ganderkesee

3. The Depths of Time
3.1 © Scala Group-Florenz
3.2 © Author
3.3 © David Haring, Duke University Primate Center, Durham, N.C.
3.4 Vera Kassühlke (after Kappelman et al. 1996)
3.5 Vera Kassühlke (after Koch et al. 2003)

4. Europe in the Eocene
4.1 Vera Kassühlke (after Koenigswald & Storch 1998: 14)
4.2 Vera Kassühlke (design by the author)
4.3 © Author
4.4 © Forschungsinstitut und Naturmuseum Senckenberg Frankfurt, photo: Erwin Haupt
4.5 © Author
4.6 After Harms et al. 2003.

5. The Dawn Horses of the Morning Cloud
5.1 Vera Kassühlke
5.2 © Author
5.3 Drawing by R. Dobrick
5.4 © Author, Geiseltalmuseum Halle Archives
5.5 Vera Kassühlke (after Franzen 1995)
5.6 © Author
5.7 © Author (Used by permission of Institut Royal des Sciences Naturelles de Belgique, Brussels)
5.8 © Forschungsinstitut Senckenberg, photo: Sven Tränkner
5.9 © Forschungsinstitut Senckenberg, photo: Sven Tränkner
5.10 © Author
5.11 © Author (Used by permission of Mensch und Natur Museum, Munich)
5.12 © Forschungsinstitut Senckenberg, SEM photo: Dr. Gotthard Richter
5.13 © Author
5.14 © Forschungsinstitut Senckenberg, photo: Dr. Friedemann Schaarschmidt
5.15 © Forschungsinstitut Senckenberg, SEM photo: Dr. Gotthard Richter
5.16 © Hessisches Landesmuseum Darmstadt, photos: Wolfgang Fuhrmannek and Marisa Blume
5.17 © Author
5.18 © Forschungsinstitut Senckenberg, photo: Sven Tränkner
5.19 © Forschungsinstitut Senckenberg, x-ray: Dr. Jörg Habersetzer and Anika Hebs
5.20 © Author (Used by permission of Mensch und Natur Museum, Munich)
5.21 © Naturkundemuseum Dortmund, photo: Dr. Walter Tanke
5.22 © Author
5.23 © Forschungsinstitut Senckenberg, photo: Sven Tränkner
5.24 Vera Kassühlke (after Nickel, Schummer, & Seiferle, Anatomie der Haustiere, 2nd ed., vol. 1, pp. 85, 161–162; Paul Parey 1961)
5.25 © Forschungsinstitut Senckenberg, photo: Sven Tränkner
5.26 Vera Kassühlke (after Schaarschmidt/Helfricht from Schaal & Ziegler, 1988: Fig. 85)
5.27 Pavel Major, Prague
5.28 Pavel Major, Prague
5.29 © Author
5.30 Vera Kassühlke (design by the author)
5.31 © Forschungsinstitut Senckenberg, SEM photo: Author and Dr. Dieter Fiege
5.32 © Naturhistorisches Museum Mainz/ Landessammlung für Naturkunde Rheinland-Pfalz
5.33 © Naturhistorisches Museum Mainz/ Landessammlung für Naturkunde Rheinland-Pfalz
5.34 a: © Forschungsinstitut Senckenberg, photo: Elke Pantak-Wein
5.34 b: © Author
5.35 © Forschungsinstitut Senckenberg, SEM photo: Author and Marie-Luise Tritz
5.36 © Naturhistorisches Museum Mainz/ Landessammlung für Naturkunde Rheinland-Pfalz

6. Constructions and Functions
6.1 Vera Kassühlke (after Thenius & Hofer 1960)
6.2 Photo: author; graphic: Vera Kassühlke
6.3 Vera Kassühlke (after Kummer 1965 and Napier 1967)
6.4 Vera Kassühlke (after Nickel, Schummer, & Seiferle, Anatomie der Haustiere, 2nd ed., vol. 1: Table 10, Fig. 494; Paul Parey 1961)
6.5 Vera Kassühlke (after Simpson 1977: 194–195: Fig. 30–31)
6.6 Vera Kassühlke (after Simpson 1977: 202, Fig. 34)
6.7 © Author (after Camp & Smith 1942)
6.8 © Nature Picture Library Bristol, UK; photo: Jim Clare
6.9 Vera Kassühlke (after Camp & Smith 1942)
6.10 Vera Kassühlke (design by the author after Tobien 1959: panel 25)
6.11 © Kit Houghton photography, Bridgwater, Somerset, UK
6.12 a: © Author
6.12 b: © Tiergarten Nürnberg, photo: Dr. Helmut Mägdefrau
6.13 Vera Kassühlke (after Otto Garraux from Bernor et al. 1997)
6.14 Vera Kassühlke (after Simpson 1977: 192, Fig. 29)
6.15 Vera Kassühlke (after Renders & Sondaar (1987), Clarendon Press, 477, Fig. 12.17, Plate 12.16)
6.16 Vera Kassühlke (after Hermanson & Hurley 1990, also Hildebrand 1987 from MacFadden 1992: 257–258, Fig. 11.18, 11.19)
6.17 Vera Kassühlke (after MacFadden 1992: 237, Fig. 11.4)
6.18 Vera Kassühlke (after Quade et al. 1989 from MacFadden 1992: 244, Fig. 11.11)
6.19 Vera Kassühlke (after MacFadden 1992: 185, Fig. 8.9)

7. Discovering Horse Evolution
7.1 © Author (from Owen 1840)
7.2 Vera Kassühlke (after Franzen 1984: Fig. 2)
7.3 © Author (from Franzen 1984: Fig. 5)

7.4 © American Museum of Natural History New York, photo: Thompson 1903
7.5 © American Museum of Natural History New York, photo: Thompson 1903
7.6 © U.S. Geological Survey (from Osborn 1929: The Titanotheres)
7.7 © American Museum of Natural History, New York, photo: Irving 1895
7.8 Vera Kassühlke (after Edinger 1948 and Radinsky 1976)
7.9 © Deutsches Pferdemuseum Verden (Aller); Punktierstich: Whessel, after John N. Sartorius 1800; photo: Michael Hensel
7.10 © American Museum of Natural History New York, photo: Eadweard Muybridge
7.11 © Blackwell-Verlag Berlin (from Simpson 1977: 122, Fig. 13)

8. Evolution and Expansion of the Horses
8.1 Vera Kassühlke
8.2 © Dr. Yuanqing Wang, Institute of Vertebrate Paleontology and Paleoanthropology, Chinese Academy of Sciences, Beijing, Peoples Republic of China, photo: Xun Jin
8.3 Vera Kassühlke (design by the author)
8.4 © Author (after MacFadden 1992: 55, Fig. 4.4)
8.5 Vera Kassühlke (after Osborn 1929)
8.6 © Author
8.7 © Kenneth D. Rose, Johns Hopkins University, photo: Shawn Zack
8.8 © Author
8.9 Vera Kassühlke (after Gingerich 2003: 473, Fig. 6)
8.10 Vera Kassühlke (after Gingerich 1989: 61, Fig. 40)
8.11 © National Park Service, photo: Arvid Aase, Fossil Butte National Monument, Wyoming
8.12 © Staatliches Museum für Naturkunde Karlsruhe, photo: Volker Griener
8.13 © American Museum of Natural History, New York
8.14 © Author
8.15 © Author
8.16 Vera Kassühlke (after Abusch-Siewert 1983: 319, Fig. 105)
8.17 Vera Kassühlke (after Abusch-Siewert 1983: 17, Fig. 3, and 85, Fig. 19b)
8.18 © Author
8.19 © Author (from Klipstein & Kaup 1836)
8.20 Vera Kassühlke (modified from Thenius & Hofer 1960: 213, Fig. 43)

9. Pseudo Horses and Relatives of Horses
9.1 Vera Kassühlke (after Janis 1976, Fig. 3)
9.2 Vera Kassühlke (after Janis 1976, Fig. 4)
9.3 Vera Kassühlke (after Osborn 1929)
9.4 © Dr. Loic Costeur Author (Naturhistorisches Museum Basel)
9.5 © Senckenberg-Museum Frankfurt, photo: Sven Tränkner
9.6 Vera Kassühlke (after Prothero & Schoch 2002, 251: Fig. 13.12)
9.7 Vera Kassühlke (after Thenius & Hofer 1960, 198: Fig. 39)
9.8 © M-O and J Plassard (Cliché no. 183 [841])
9.9 © Verlag Ferdinand Berger & Sons, Horn, Österreich (from Zapfe 1979, 270: Fig. 155)
9.10 © Author (Naturhistorisches Museum Basel)
9.11 © U.S. Geological Survey (from Osborn 1929)
9.12 Pavel Major, Prague
9.13 Vera Kassühlke (after Scott 1930, from Prothero & Schoch 2002, 215: Fig. 11.3A)
9.14 Vera Kassühlke (after Colbert 1991, from Prothero & Schoch 2002, 215: Fig. 11.3B, corrected)

10. The Ice Age and the Roots of Modern Horses
10.1 Vera Kassühlke (after Berner & Streif 2000: 55, Fig. 3.14)
10.2 Vera Kassühlke (after Berner & Streif 2000: 18, Fig. 2.5)
10.3 Vera Kassühlke (after Berner & Streif 2000: 221, Fig. 11.15)
10.4 © Author (Used by permission of Staatlichen Museums für Naturkunde Stuttgart)
10.5 Vera Kassühlke (after MacFadden 1992: 111, Fig. 5.20)
10.6 Vera Kassühlke (after Eisenmann, in press)
10.7 © Author (Naturhistorisches Museum Basel)
10.8 Vera Kassühlke (from a design by the author together with Véra Eisenmann, Muséum National d'Histoire Naturelle Paris)
10.9 © Naturhistorisches Museum Mainz Archives, photo: Author ch14_Abbildungsnachweis_klein.qxd 21.07.2006 11:51 Uhr Seite 210
10.10 © Ministerium der Republik Frankreich für Kultur und Kommunikation, Region Rhône-Alpes, Abteilung für Archäologie
10.11 © Véra Eisenmann, Muséum National d'Histoire Naturelle Paris

11. Conclusion
11.1 Vera Kassühlke (design by the author after Simpson 1977, Thenius 1969, MacFadden 1992, and the author's own research)

Exhibits on the Evolution of Horses
p. 186: *top*, Florida Museum of Natural History, photo: Mary Warrick © 2008
bottom, © Hessisches Landesmuseum Darmstadt (Used by permission of Hessisches Landesmuseum Darmstadt)
p. 187: *top*, © Author (Used by permission of Naturkundemuseums Dortmund)
bottom, © Author (Forschungsinstitut und Naturmuseum Senckenberg Frankfurt)
p. 188: *top*, © Geiseltalmuseum/Halle
bottom, © Niedersächsisches Landesmuseum Hannover, photo: Dr. Michael Schmitz
p. 189: *top*, © Author (Used by permission of Staatlichen Museums für Naturkunde Karlsruhe)
bottom, © Author (Used by permission of Naturhistorischen Museums Mainz)
p. 190: *top*, © Maarmuseum Manderscheid, photo: K. Steinhausen
bottom, © Author (Used by permission of Museumsvereins in Messel)
p. 191: *top*, © Author
bottom, © Landau & Kindelbacher, Munich
p. 192: *top*, © Westfälisches Pferdemuseum Münster, photo: Dr. Christoph Esch
bottom, American Museum of Natural History/Craig Chesek
p. 193: *top*, © Author (Used by permission of Staatlichen Museums für Naturkunde Stuttgart)
bottom, © Deutsches Pferdemuseum Verden, photo: Michael Hensel